Mass Transfer Dynamics of Contaminants in Fractured Media

Zhi Dou · Zhifang Zhou · Jinguo Wang ·
Yong Huang

Mass Transfer Dynamics of Contaminants in Fractured Media

Science Press
Beijing

Springer

Zhi Dou
School of Earth Sciences and Engineering
Hohai University
Nanjing, China

Zhifang Zhou
School of Earth Sciences and Engineering
Hohai University
Nanjing, China

Jinguo Wang
School of Earth Sciences and Engineering
Hohai University
Nanjing, China

Yong Huang
School of Earth Sciences and Engineering
Hohai University
Nanjing, China

ISBN 978-981-99-9186-0 ISBN 978-981-99-9187-7 (eBook)
https://doi.org/10.1007/978-981-99-9187-7

Jointly published with Science Press
The print edition is not for sale in China (Mainland). Customers from China (Mainland) please order the print book from: Science Press.
ISBN of the Co-Publisher's edition: 978-7-03-072951-4

This Springer imprint is published by the registered company Springer Nature Singapore Pte Ltd.
The registered company address is: 152 Beach Road, #21-01/04 Gateway East, Singapore 189721, Singapore

Preface

In recent years, with the rapid development of social economy and the continuous improvement of the level of industrialization, the problem of soil and groundwater pollution has attracted great attention. According to incomplete statistics, there are more than 500 thousand heavily polluted sites in China due to the relocation and abandonment of factories. Since the 1970s, developed countries have researched the spatial distribution and transport of dissolved organic pollutants in soil and groundwater, especially the multiphase flow, the process of multifactorial pollutant transport, and the prediction and simulation at different scales. In China, there are few researches in this field, and there is a lack of systematic theoretical system and practical experience. Due to the coupling of multiple scales and domains, multiphase and multicomponent processes, and other characteristics of mass transfer of pollutants in aqueous media, the problems of mass transfer are interdisciplinary fields for hydromechanics, groundwater dynamics, reaction kinetics, soil science, microbiology, etc. The kinetics of mass transfer of pollutants in aqueous media is a very comprehensive research field and a direction that needs to be developed rapidly and be further developed to analyze the causes of pollution and control technologies. Therefore, there is still much room for improvement in both theory and practice. This textbook on the dynamics of mass transfer of pollutants in fractured media was written to focus on the mass transfer of pollutants and the dynamic process of mass transfer of pollutants at multiphase interfaces. It is not only the core system for analyzing the mechanism of multiphase mass transfer of pollutants in complex fractured aqueous media, but also an important basic theory for solving important scientific problems in the field of groundwater pollution control and treatment.

The textbook draws on the national strategic needs of building an ecological civilization, relies on the national key project of the research and development program of soil pollution causes and control technologies, integrates the inherent needs of scientific research and education, and absorbs the latest global research results of groundwater pollution mass transfer kinetics. Based on the requirement of cultivating the knowledge structure of highly skilled creative talents, this textbook emphasizes theoretical, systematic, boundary-spanning, and holistic content presentation. At the same time, the textbook highlights the most important points, covers comprehensive knowledge, and systematically integrates specialized knowledge with Chinese and English expressions. Based on the basic vocabulary, key sentence patterns, and literature reading, this textbook will improve students' English language use, so that literature reading and English writing skills will provide a solid foundation for students to advance their research work.

This textbook integrates political education throughout the writing process so that students can properly understand the urgency and necessity of groundwater protection in China. For example, by analyzing the causes of groundwater pollution in typical locations, the green concept of China's sustainable development is properly explained, and the relationship between ecological environmental protection and management, and geological environmental protection in the context of the living community of all living things is clarified to improve cultural and strategic confidence. Based on the similar textbooks available in the field, we strengthen the knowledge points of groundwater pollution shift theory and technology in complex areas based on the main contents of traditional groundwater dynamics and fracture media hydrodynamics. At the same time, the textbook compiles and cites the review articles published in prestigious journals in recent years. The emphasis of the textbook is on deepening the kinetic theory of contaminant transfer in groundwater in fractured media, in order to consolidate students' basic knowledge of kinetic analysis and to expand their knowledge reserves.

We thank Wenyuan Zhu, Yan Zhao, Ze Yang, and Yufeng Huang for their careful editing. We also acknowledge the support from the National Key Research and Development Program of China (Grant No. 2019YFC1804303).

Nanjing, China

Zhi Dou
Zhifang Zhou
Jinguo Wang
Yong Huang

Contents

About the Authors

Prof. Zhi Dou is a full-time professor at Hohai University. He conducts research and teaches in the fields of engineering geology, hydrogeology and environment, groundwater pollution and control. He has led more than 10 research projects, including the National Key Research and Development Program and the National Natural Science Foundation of China. He has also done consulting research work for several major national projects in water conservancy and hydropower.

Prof. Zhifang Zhou is a full-time professor at Hohai University. He teaches and conducts research in the field of hydrogeology and engineering geology. His research interests include hydrogeological condition assessment, groundwater pollution and prevention, rock (soil) engineering and environment, etc. He published 8 monographs, textbooks, and industry regulations and published more than 150 scientific papers.

Prof. Jinguo Wang is a full-time professor at Hohai University. His research interests include the study of hydrogeological conditions in water resources management and hydropower engineering, hydrogeological parameter acquisition technology, karst groundwater simulation and prediction, and groundwater pollution control. He has published over 130 scientific papers and two monographs, participated in the drafting of three industry regulations, and has 15 invention patents approved.

Prof. Yong Huang is a full-time professor at Hohai University. His research interests involve the study of groundwater issues in engineering, simulation of groundwater flow and solute transport, evaluation of groundwater resources, and inversion simulation of hydrogeological parameters, etc. He has published over 100 scientific papers and one monographs and has 8 invention patents approved.

Frequently Used Symbols

A	Area (L^2)
A_0	The sum of bottom area (L^2)
A_t	The fracture contact area (L^2)
b	The aperture within a single fracture (L)
$\bar{b}$	The mean aperture between top and bottom walls (L)
C_m	Solute concentrations in the mobile domains (ML^{-3})
C_{im}	Solute concentrations in the immobile domains (ML^{-3})
C	Solute concentration (ML^{-3})
D	The dispersion coefficient (L^2T^{-1})
$D_{f, mim}$	The fitted dispersion coefficient (L^2T^{-1})
d_0	The fracture diameter (L)
D_m	The molecular diffusion coefficient (L^2T^{-1})
D_{macro}	The macro-dispersion (L^2T^{-1})
d_0	The shear displacement distance along the horizontal direction (L)
D_0	Free solution diffusion coefficient (L^2T^{-1})
D_{ADE}	Effective dispersion of ADE model (L^2T^{-1})
D_{CTRW}	Dispersion in the framework of the CTRW (L^2T^{-1})
$D_{f, m}$	Dispersion coefficient (L^2T^{-1})
D_L	Coefficients of longitudinal dispersion (L^2T^{-1})
D_T	Coefficients of transverse dispersion (L^2T^{-1})
D_{Taylor}	Taylor dispersion coefficient (L^2T^{-1})
D_x	Molecular diffusion coefficient in x-direction (L^2T^{-1})
D_y	Molecular diffusion coefficient in y-direction (L^2T^{-1})
D_z	Molecular diffusion coefficient in z-direction (L^2T^{-1})
e	The largest local aperture (L)
e_h	Hydraulic aperture (L)
F_{st}	The surface tension force acting at the interface between two immiscible fluids ($ML^{-2}T^{-2}$)
g	Acceleration of gravity (LT^{-2})
G	Field intensity gradient (–)
H	Continuous plane-source of height (L)

h	The imbibition height (L)
h_{max}	The maximum height of the capillary imbibed water (L)
h_1	Inlet water head (L)
h_2	Outlet water head (L)
h_x	The horizontal distance (L)
i	Hydraulic gradient (–)
K	Hydraulic conductivity (LT^{-1})
k	Intrinsic permeability (L^2)
K_w	The effective hydraulic conductivity (LT^{-1})
K_f	Permeability coefficient of fracture surfaces (LT^{-1})
L	The length of the flume (L)
l	Distance in the direction of transfer (L)
L_z	Distance of solute transport (L)
m	Fluid velocity (LT^{-1})
M	Total mass (M)
m_0	The mass of injected solute (M)
n	Porosity (–)
n_e	Effective porosity (–)
p	The fluid pressure ($ML^{-1}T^{-2}$)
p_{nw}	Capillary force in non-wetting phase ($ML^{-1}T^{-2}$)
P_{th}	The capillary pressure ($ML^{-1}T^{-2}$)
p_w	Capillary force in wetting phase ($ML^{-1}T^{-2}$)
p_c	Capillary force ($ML^{-1}T^{-2}$)
Q	The total discharge rate (L^3T^{-1})
q	Seepage velocity (LT^{-1})
Q	The flow rate (L^3T^{-1})
Q_w	The volume of the imbibed water in the matrix (L^3)
q_{im}	Porosity in the immobile domains (–)
q_m	Porosity in the mobile domains (–)
R	Pore radius (L)
r_e	Radius of influence (L)
$\bar{r}$	The average radius of the grains (L)
r_w	Well radius (L)
r_1, r_2	The radius of curvature of the two-phase interface (L)
Re_c	The critical Reynolds number (–)
S_w	Wetting phase (–)
S_{nw}	The saturation of non-wetting phase (–)
S_{nw0}	Residual saturation of the non-wetting water phase (–)
S_{w0}	Irreducible saturation of the water phase (–)
t	Time (T)
T_2	The transverse relaxation time (T)
T_{2b}	Bulk relaxation (T)
T_{2d}	Diffusion relaxation (T)
T_{2s}	Surface relaxation (T)
T_k	Temperature (–)

u	The flow velocity (LT^{-1})
$\bar{u}$	Average flow velocity (LT^{-1})
u_{av}	The average velocity magnitude in the porous media (ML^{-1})
$\bar{u}_m$	The effective flow velocity in mobile domain for the wetting phase (ML^{-1})
u_y	Transport velocity in the framework of the CTRW (LT^{-1})
V	Volume (L^3)
v	The mean velocity of the flow (LT^{-1})
W	The width of fracture in the out of plane direction (L)
w_i	Weights for the D2Q9 model (M)
$X(t)$	Random motion (L)
α	Mass transfer rate between the mobile and immobile domains (T^{-1})
α_L	Longitudinal dispersion (L)
α_T	Transverse dispersion (L)
α_x	Longitudinal dispersivity (L)
α_y	Horizontal-transverse dispersivity (L)
α_z	Vertical-transverse dispersivity (L)
γ	The mobility parameter ($M^{-1}L^3T$)
ΔV_n	Volume of space (L^3)
ΔV_α	Volume of fluid α (L^3)
Δx	Solute molecule movement (L)
ϵ	The thickness of the interface between the two immiscible fluids (L)
η	Fluid viscosity (L^2T^{-1})
λ	The magnitude of the mixing energy (MLT^{-2})
γ	The weight of the water (ML^{-3})
λ_T	The equivalent capillary diameter (L)
μ	Dynamic viscosity (L^2T^{-1})
ρ	Density (ML^{-3})
ρ_2	Relaxation rate (–)
$\rho\sigma$	Mass densities (ML^{-3})
σ	The surface tension coefficient (MT^{-2})
τ	The tortuosity for diffusion in fractures (–)
τ^σ	The relaxation time (T)
Φ	The porosity of the medium (–)
φ	The porous medium porosity (–)
ω_i	Weight (M)

Chapter 1
Introduction

NAPL (Non-aqueous Phase Liquid) is a typical subsurface organic pollutant, such as tetrachloroethylene, trichloroethylene, and methylbenzene. Although the solubility of NAPLs in water is quite low (for example, the solubility of tetrachloroethylene in water is only 0.015 g/100 mL), their harmfulness to the groundwater environment is surprising. Many NAPLs pose a high risk of cancer. The World Health Organization's drinking water standard requires that benzene and trichloroethylene contain no more than 0.005 mg per liter of water, and vinyl chloride, which poses a higher cancer risk, must contain no more than 0.002 mg per liter of water. Although the solubility of many NAPLs in groundwater is quite low, it is still above the safety standard for drinking water. NAPLs are divided by the density of pure water. One is heavier than water and is referred to as DNAPL (dense NAPL), while another is lighter than water and is referred to as LNAPL (light NAPL). Compared to the LNAPL, the DNAPL has a higher density than water and sinks continuously after entering the groundwater.

DNAPL has the properties of concealment and persistence, and the problems of migration and mass transfer in fractured groundwater are more complex. DNAPL is a very specific organic contaminant. Its migration and mass transfer process has the characteristics of soluble and insoluble pollutants. This is mainly because most liquid DNAPL cannot be completely dissolved in a short time and discharged from the subsurface with the runoff, but remain in the frac medium for a long time in the form of a "DNAPL pollution pool" that slowly dissolves and expands its plume with the runoff from the subsurface. The process of DNAPL migrating with the water flow as a liquid in a non-aqueous phase and DNAPL migrating with the water flow after being dissolved from the water is referred to here as mass transfer. The mass transfer process of DNAPL in fractured groundwater consists of the migration of an immiscible two-phase stream and the migration of solutes after the DNAPL is dissolved.

Mass transfer of contaminants in fractured media plays an important role in surface environmental problems (e.g., contaminant degradation (Sale et al. 2008), nuclear waste disposal (Bagalkot and Kumar 2016), and bioremediation (Song et al. 2014). In

Z. Dou et al., *Mass Transfer Dynamics of Contaminants in Fractured Media*,
https://doi.org/10.1007/978-981-99-9187-7_1

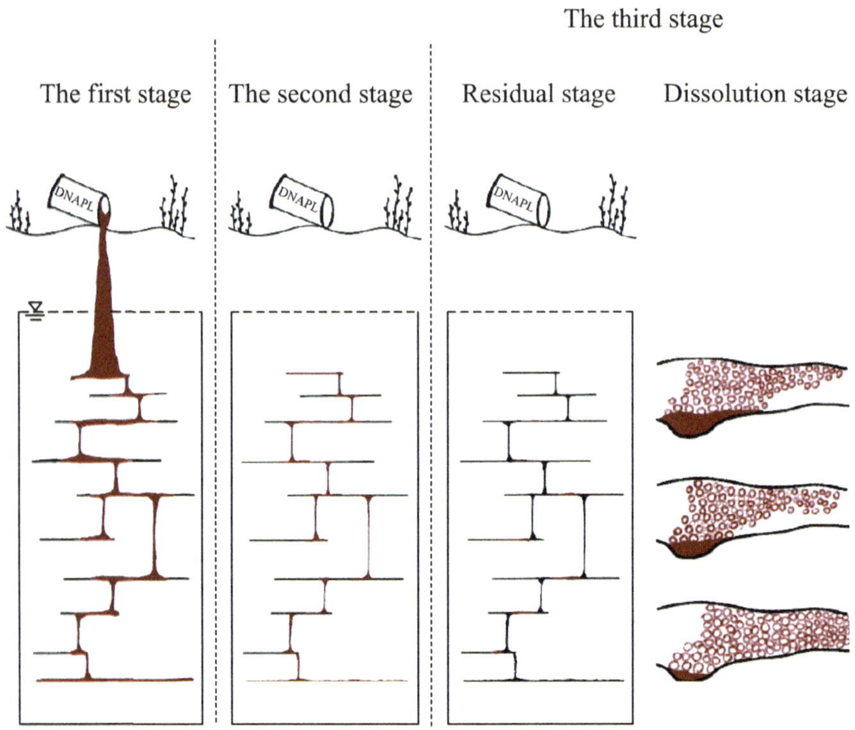

Fig. 1.1 Schematic diagram of DNAPL invasion process

recent decades, the study of dilution and mixing processes in fractured media has been focused on by scientists from various fields (Anna et al. 2014; Soltanian et al. 2015; Dou et al. 2018a, 2018b). Characterization of the dispersal and mixing processes of the conservative solute is critical to understanding and evaluating reactive solute transport, which is necessary for studying complex chemical-biological reaction transport in groundwater. Although many studies (Shapiro and Brenner 1988; Pini et al. 2016) have provided new insights into the mechanisms and properties of mixing behavior at the Darcy scale or the larger field scale, little attention has been paid to porous media at the pore scale. There is always an influence of the smaller scale processes on the larger scale behavior (Bear 2013). Therefore, it is of great importance to understand and evaluate mixing processes at the pore scale using appropriate methods (Rolle et al. 2013).

The migration process of DNAPL in fractured groundwater can be divided into three phases: The first phase is the general intrusion phase of DNAPL; the second phase is the comigration phase of DNAPL and water flow; the third phase is the residual dissolution phase of DNAPL, as shown in Fig. 1.1.

The migration process of DNAPL in fractured groundwater can be divided into three phases: The first phase is the general intrusion phase of DNAPL; the second phase is the comigration phase of DNAPL and water flow; the third phase is the residual dissolution phase of DNAPL, as shown in Fig. 1.1. The residual DNAPL of fractured groundwater refers to the DNAPL that cannot be displaced by water flow alone under conditions of natural water flow and pressure. Since the dissolution rate of DNAPL is affected by its interface area, hydrodynamic conditions, and geometric features of fractures, its dissolution content in water can be neglected compared to the penetration of DNAPL in a short time.

In general, the migration process of DNAPL in the first and second stages is considered only as a two-phase immiscible flux migration process, while the solubility of DNAPL in water is ignored. In the third stage, the remaining DNAPL is relatively stable and no longer migrates as a whole, but continuously dissolves in the water body and increases its dissolved contaminant plume due to the effect of groundwater discharge. At this stage, the problem of DNAPL dissolution process and solute transport with DNAPL as solute must be considered. The first and second stages directly affect the distribution characteristics of DNAPL residue in fractured media in the third stage, and the third stage is an important reason for the persistence and concealment of DNAPL pollutants.

The spatial heterogeneity of pore space in aquifers results in complex fluid flow and anomalous solute transport. The occurrence of preferential flow paths and/or cementations found in different systems results in non-Fickian solute transport. Cementation of solid grains is ubiquitous in natural soils and is one of the most common causes leading to changes in pore space and heterogeneity of the porous medium. In many laboratory and field scale studies (Fiori et al. 2006; Neuman and Tartakovsky 2009; Bijeljic et al. 2013a; Swanson et al. 2015), the experimental results showed that solute transport follows non-Fickian transport. Some studies (Edery et al. 2014; Heidari and Li 2014; Hou et al. 2018) indicated a strong relationship between non-Fickian transport and heterogeneity of porous media by fitting the experimentally measured or directly simulated solvent plume breakthrough curves with the continuous-time random walk (CTRW) models. The different geometry of pore channels and cementation of solid grains are very common in natural solids, leading to heterogeneity of porous media. Previous studies indicated that the properties of pore channels, such as connectivity or pore particle distribution, can influence the flow field and mass transfer processes (Bijeljic et al. 2013b; Voller 2015). The evolution of solute mixing in porous media with different degrees of heterogeneity is caused by the changes in the geometry of pore channels and solid grains due to the cementation of porous media (Hou et al. 2018). However, little attention has been paid to this problem.

For the heterogeneous system, the flow velocity distribution in the porous media and fractured media is a fundamental issue for understanding the mixing behavior of solute plume (Gramling et al. 2002). As the mixing process is highly influenced by the flow field (Rolle et al. 2012; Chiogna et al. 2015), molecular diffusion causes the solute molecule movement between streamlines and enhances mixing. Some recent studies characterized topological (Barros et al. 2012) and kinematic (Le Borgne et al. 2013) feature of the flow field to explore the influence of the flow field on the mixing behavior of the solute plume.

In many contaminant transport studies, insufficient mixing could limit mass conversion or rates of contaminant conversion. Conservative solute transport could be viewed as a process composed of dispersion and mixing. However, the difference between dispersion and mixing must be distinguished in detail (Dou et al. 2016). Spreading is the change in the shape of the solvent plume associated with the deformation and elongation caused by the variability of the flow field. The mixing indicates the redistribution of concentration and tends to be uniform within the solvent plume. Many approaches have been proposed to quantify and describe the mixing behavior. In homogeneous porous media, the second central spatial moments can be adopted as a good measure for quantifying the mixing.

However, it cannot accurately reflect the mixing (or dilution) processes in a heterogeneous system (Rolle et al. 2009). Kitanidis (1994) proposed the dilution index, which measures the volume occupied by the solution plume. Rolle et al. (2009) proposed a flow-based dilution index based on the concept of dilution index to quantify dilution in a steady-state flow with continuous injection. Kapoor and Kitanidis (1996) used concentration variance to quantify mixing. Several researchers quantified mixing in subsurface problems by the scalar dissipation rate (SDR), which was originally proposed to describe turbulent flow and combustion. Pore-scale mixing is a necessary condition for the reaction between different solute compounds (Kitanidis 1994; Cirpka 2002). Therefore, quantification of pore-scale mixing is key to understanding and predicting contaminant transport or reaction.

1.1 Mass Transfer in Saturated System

It is generally accepted that fractures can play an important role in the transport and fate of contaminants. Characterizing the dispersion and mixing processes of conservative solutes through fractures is very important for understanding reaction and mass transport rates associated with nuclear waste disposal, enhanced oil recovery, and bioremediation (Berkowitz et al. 2006; Dentz et al. 2011; Cai et al. 2017). Although many studies in recent decades have provided new insights into the mechanisms and properties of mixing processes in homogeneous and heterogeneous porous media (Kitanidis 1994; Kapoor et al. 1998; Cirpka and Kitanidis 2000; Dentz et al. 2004; Dentz and Carrera 2007; Rolle et al. 2009; Tartakovsky et al. 2009; Le Borgne et al. 2010; Chiogna and Bellin 2013; Dentz et al. 2015; Chiogna et al. 2016), little attention has been paid to mixing behavior in fractures.

Since the heterogeneity of geologic formations is ubiquitous, a fundamental question about the difference between spreading and mixing processes of conservative solutes must be understood. Several authors (Kitanidis 1994; Cirpka 2002; Dentz and Carrera 2007) have emphasized the difference between spreading and mixing. Dispersal refers to the change in the spatial extent of a contaminant plume, while mixing describes the process that unifies the concentration distribution of contaminants within the plume. In other words, spreading results in expansion and deformation of a solvent plume, while mixing results in dilution of a conservative solvent over time. Thus, spreading and mixing are not the same thing, but a complete conservative solute transport can be considered as a combination of spreading and mixing.

In a conservative solute transport in a spatially variable velocity field, the spreading of the solution plume is driven by the advective forces that deform and extend the plume along the streamlines, while at the same time molecular diffusion causes the mixing that smooths the concentration gradients within the solution plume. There is a complex interaction between the spreading and mixing processes, especially in heterogeneous flow fields. Because of the natural coupling of dispersion and mixing, separating the dispersion and mixing processes is challenging, but studying the time evolution of mixing is still useful and important for improving predictions of reactive transport and mixing. The description of conservative solute transport by dispersion only is valid for some applications (such as risk analysis). However, it is inadequate when the transport problem involves mixing-driven chemical reactions in which the reaction rate is highly dependent on the mixing of reactants (de Simoni et al. 2005; de Dreuzy et al. 2012; Rolle and Kitanidis 2014).

To this end, numerous efforts have been made to develop methods for quantifying mixing and dispersal. According to the moment method of Aris (1956), the second central spatial moment of a conservative solute is a measure of dispersion. This is because the spatial extent of the plume can be easily estimated based on the time evolution of the plume's spatial moments and is related to an apparent dispersion coefficient even for preasymptotic times (Freyberg 1986; Detwiler et al. 2000; Dentz and Carrera 2007; Bolster et al. 2011). In a homogeneous flow field, the second central spatial moment increases monotonically with time, which can be expected to be a good measure of both dispersion and mixing. This approach is invalid for a heterogeneous flow field, where the second central spatial moment could decrease due to streamline convergence (Cirpka and Kitanidis 2000; Rolle et al. 2009, 2012). Several metrics have been proposed for quantifying mixing.

Since dilution caused by mixing is an irreversible process, Kitanidis (1994) proposed the dilution index, which measures the volume occupied by the dissolved plume. The dilution index is derived from the statistical entropy of the solute distribution. Unlike the second central spatial moment, the dilution index is able to quantify the actual mixing, and the mixing rate can be calculated using the rate of change of entropy. The dilution index is useful not only for conservative mass transfer but also for reactive transport (Chiogna et al. 2012). The ratio between the actual and theoretical maximum dilution index can be an indicator of the influence of incomplete mixing on reactive transport (Tartakovsky et al. 2009; Rolle et al. 2013). In addition, the flow-based dilution index, based on the original concept of dilution index proposed by Rolle et al. (2009) for steady-state transport with continuous injection, describes dilution as "the act of distributing a given mass flux of solute over a larger flux of water".

From the perspective of stochastic hydrogeology, the concentration distribution of a contaminant plume can be decomposed into a cross-sectional mean and a fluctuation around that mean. The concentration variance method has been proposed as a measure of mixing (Kapoor and Gelhar 1994; Kapoor et al. 1998; Miralles-Wilhelm and Gelhar, 2000). In addition to the dilution index and concentration variance, mixing can alternatively be quantified by the scalar dissipation rate (SDR), which is determined from the concentration gradients within the solution plume (de Simoni et al. 2005; Le Borgne et al. 2010). Although SDR has been proposed for the study of turbulent flow and combustion, several studies have shown that SDR can also be applied to a variety of subsurface contaminant transport problems (e.g., compound-specific transport, conservative solute transport, and reactive multicomponent transport). Le Borgne et al. (2010) studied the time evolution of SDR in heterogeneous porous media and demonstrated the occurrence of non-Fickian scaling of mixing. Bolster et al. (2011) used SDR to split the global mixing state into a dispersive mixing state and a local mixing state. Jha et al. (2011) used SDR to quantify mixing in a viscosity-unstable flow.

de Dreuzy et al. (2012) argued that mixing results from competition between velocity fluctuations and local diffusion, and proposed a new decomposition of mixing into potential mixing and churn rate. This new decomposition of mixing showed a generic characterization and could provide new ways to establish a transport equation considering advection, dispersion, and mixing. A comparison of the different transport models can be found in (de Dreuzy and Carrera 2016). In addition, a number of analytical solutions for SDR in non-conservative transport systems have been derived by Engdahl et al. (2013). However, these current studies of SDR are limited to specific homogeneous or heterogeneous porous media. Because anomalous (non-fissure) transport has been observed in individual rough fractures (Cardenas et al. 2007, 2009; Wang and Cardenas 2014, 2015; Kang et al. 2016) and mixing and spreading in fractures may play an important role, investigation of the performance, characteristics, and evolution of SDR in other major subsurface geologic formations (e.g., individual rough fractures) needs to be further explored.

Fissures in geologic formations with low permeability represent the most important flow paths. Characterization of solute transport through fractures is critical to understanding and quantitatively assessing environmental and engineering problems. In practice, fracture deposits consist of numerous individual fractures and form complex fracture networks in the subsurface. However, detailed characterization of the complex fracture networks remains a challenge. Many studies (Tsang et al. 1988; Cardenas et al. 2007, 2009; Boschan et al. 2008) have focused on discrete individual fractures with variable apertures that form the core of the fracture network model. Therefore, the behavior of mass transfer in the discrete single fracture with variable aperture has yet to be evaluated.

Mass transport mixing and dispersion in fractures with variable apertures are usually assumed to follow Fick's process, in which the dispersion coefficient is constant in space and time. However, this is based on the assumption that the geometry of the opening fields and the morphology of the fractures are statistically homogeneous and stationary. Under this assumption, the complex fracture is reduced to an ideal 2D parallel fracture, neglecting the rough fracture walls and aperture variations. The effective longitudinal dispersion D_L could be determined by fitting the breakthrough curves (BTCs) to the one-dimensional advection–dispersion equation. Keller et al. (1999) compared the observed BTCs at different flow velocities with the predicted BTCs using small perturbation analysis and direct numerical simulation in rough fractures. Their results showed a linear relationship between the D_L and the average velocity $\overline{u}$.

Because of the different dispersion mechanisms in fractures with variable opening, the prediction of solute transport behavior is possible by using this linear relationship between the D_L and $\overline{u}$ could be inappropriate over a wide range of dimensionless Pelect numbers ($Pe = \overline{u}\overline{b}/D_m$, where $\overline{b}$ is the arithmetic mean aperture and D_m is the molecular diffusion coefficient). In addition to the linear relationship, nonlinear relationships (i.e., power law and second-order polynomial relationships) have also been proposed for predicting solute transport behavior. Dronfield and Silliman (1993) reported a power law relationship between the D_L and $\overline{u}$ in which the exponent is equal to 1.3 in the variable-aperture fracture and equal to 2 for the ideal 2D paralleled fracture based on Taylor dispersion theory (Aris, 1956). It is worth noting that the results of the work of Dronfield and Silliman (1993) also suggest that the roughness scale may have an effect on the power law relationship based on solute transport experiment in an analogous sand quarry; however, they did not provide provable experimental data. Ippolito et al. (1994) found experimentally that, because of the roughness of the fracture, macrodispersion dominates solute dispersion at low flow rates, while Taylor dispersion dominates at high flow rates. A scaling analysis performed by Roux et al. (1998) showed that three different dispersion mechanisms (i.e., molecular diffusion, macrodispersion, and Taylor dispersion) occur in fractures with variable opening. With respect to conservative solutes, these dispersion mechanisms are independent of each other, and each of them dominates the dispersion and mixing processes depending on the different areas of the pe. As Detwiler et al. (2000) noted, macrodispersion is caused by the velocity fluctuations induced by the variability of the aperture field, whereas Taylor dispersion results from the fluctuations of the velocity distribution across the fracture aperture. Since these dispersion mechanisms are due to different causes, macro-dispersion and Taylor dispersion show different fundamental dependence on *Pe*. For example, macrodispersion shows a proportional relationship with Pe, while Taylor dispersion shows a proportional relationship with Pe^2. Detwiler et al. (2000) investigated experimentally and theoretically the relative importance of macrodispersion and Taylor dispersion. They proposed a second-order polynomial relationship between the D_L and *Pe*, and assumed that the D_L in the fracture with variable aperture is the sum of three mixing and spreading mechanisms (i.e., diffusion, macrodispersion, and Taylor dispersion).

However, simulating and predicting mass transfer in variable-opening fractures with a linear, exponential, or second-order polynomial relationship between DL and Pe remains challenging because the roughness of fracture walls leads to spatial heterogeneities of mean flow in the plane of the fracture (Tsang and Tsang 1987; Brown et al. 1998). Many studies have shown that the opening field has a significant effect on solute transport and dispersion mechanism. Bolster et al. (2009) studied the DL in 2D channels with periodically varying apertures. They found that periodic variations in the aperture lead to both a decrease and an increase in DL. Zheng et al. (2009) used Reynolds equations to solve the flow fields in the rough fractures, and numerically investigated the interactive effect of heterogeneity and anisotropy of the fracture aperture fields on the Taylor dispersion and geometric dispersion (Marco dispersion). The results showed that the geometric dispersion became more dominant with increasing coefficient of variation (COV).

Wang and Cardenas (2014) coupled the Navier–Stokes equations and the two-dimensional advection–diffusion equation to solve the problem of solute flow and transport in variable-opening fractures. They found that the heterogeneity of the fracture opening field would enhance the extent of non-Fickian (anomalous) behavior. However, fracture wall heterogeneity occurs at all levels (Berkowitz et al. 2006; Tartakovsky and Xiu 2006; Xiu and Tartakovsky 2006). Consequently, solute mixing and dispersion could appear to be scale-dependent behaviors associated with the scale dependence of the dispersion coefficient. Thus, the relationship between DL and Pe may be sensitive to the scale of roughness even for a length-dependent fracture, implying that further investigation of the influence of multiscale roughness of fracture on the relationship between DL and Pe is needed.

The influence of multiscale roughness on flow behavior has attracted much attention in the last decade (Brown et al. 1995; Oron and Berkowitz 1998; Cardenas et al. 2009; Wang et al. 2015). Zimmerman et al. (1991) studied the influence of small-scale roughness on the transmissivity of fractures and proposed analytical expressions for the deviations from the predictions of the classical cubic law. Recently, the influences of multi-scale roughness on nonlinear fluid flow in 2D and 3D rough fractures were confirmed (Zou et al. 2015; Wang et al. 2016).

For a self-affine variable-aperture fracture wall, there exists two-scale roughnesses or wavinesses. Zou et al. (2015) used a wavelet analysis technique to decompose the original rough fracture wall into a large-scale (high-frequency) roughness and small-scale (low-frequency) roughness. The numerical simulation results (Zou et al. 2015; Wang et al. 2016) showed that the large-scale roughness controlled the pressure distribution and dynamic evolution of eddy flow regions at a large scale, while the small-scale roughness determined the nonlinear properties of fluid flow at a local scale.

Reaction transport as a chemically active process dominates the fate of groundwater contamination by fractured rock. Characterization of this fundamental transport process is central not only to groundwater remediation, but also to many subsurface natural and engineered systems, such as enhanced oil recovery, geothermal energy development, and electrical conductivity of geologic formations (Cai et al. 2017; Jin et al. 2017b). As a classical macroscopic-scale approach, the advection–dispersion-reaction equation (ADRE), which is based on the assumption of perfect mixing, usually overestimates the reaction rate and tends to make less reliable predictions. Many studies (Cao and Kitanidis 1998; Kapoor et al. 1998; Raje and Kapoor 2000; Cirpka and Valocchi 2007; Chiogna and Bellin 2013; Dentz et al. 2015; Qian et al. 2015) have shown that the ADRE is generally unable to describe the local mixing phenomenon and that neglected imperfect mixing at the pore scale is a major reason for the overestimated reaction rate. Therefore, the mixing-induced reaction transport through the fracture at the pore scale still needs to be better understood.

Since solute transport behavior is inextricably linked to fluid flow, the literature on fluid flow in rough fractures was briefly reviewed for completeness. The influence of fracture wall roughness on fluid flow has attracted much attention in recent decades (Tsang 1984; Zimmerman and Bodvarsson 1996). Based on the model of an ideal smooth parallel plate, the classical cubic law, the analytical solution of the Reynolds equation derived from the linearization of the Navier–Stokes equation, is often used to simplify fluid flow in single rough fractures. However, many numerical and experimental studies (Brush and Thomson 2003; Wang and Cardenas 2015) have found that the classical cubic law is valid only when the flow is laminar, and can lead to a non-negligible error when the fracture wall is rough. The assumption of an ideal smooth parallel plate model at the macroscopic level is the main factor that leads to the fact that the cubic law may be inadequate for the description of fluid flow and reaction transport. Therefore, the flow field in the individual self-affine fractures was solved directly by the Navier–Stokes equation at the pore scale.

The traditional macroscopic reaction transport models are based on upscaling the transport equation and the effective transport parameters. The usual procedure is to determine the effective transport parameters by fitting the model results to the flow-averaged breakthrough curves (BTCs) of a passive tracer. Because of the heterogeneity of the geologic formation, the BTCs may be non-Gaussian and have non-Fickian characteristics (i.e., "early arrival" and "strong runout"). Although using the macroscopic advection–dispersion equation (ADE) to fit the non-Fickian BTCs could result in a non-negligible error, the non-Fickian BTCs can provide important information for understanding the mixing process. For example, the "early arrival" in the BTCs indicates that the preferred pathway is in fractures, whereas the "heavy tail" indicates that a slow mass transfer process controls transport at late times (Haggerty et al. 2000). Therefore, the non-fissure BTCs with "heavy tail" can be expected to serve as indicators of mixing processes in heterogeneous geologic formations.

Luo and Cirpka (2011) numerically investigated the validity of macroscopic transport models, capable of describing all the details of BTCs, for predicting a reaction transport in heterogeneous media. They found that the direct application of the macroscopic transport model to the reaction transport is inappropriate due to the concentration fluctuations present. Moreover, the mixing process, which is different from the spreading process, represents the distribution of a solute over an increasingly larger volume. This is the only process that allows mass transfer between different streamlines and results in a decrease in peak concentration. Intermixing can also be considered the process that blurs concentration differences. The process of reactant mixing is critical for accurate prediction of mixing-induced reaction transport. However, the influence of vortices on the mixing process in a self-affine fracture is still an open issue.

1.2 Mass Transfer in Unsaturated System

Transport of aqueous tracers in unsaturated systems (i.e., at least immiscible fluids) occurs in natural processes, industrial applications, and innovative tracer technologies, such as sustainable unconventional irrigation processes, enhanced oil recovery, geothermal energy development, organic contaminated groundwater remediation, interfacial distribution tracer testing, and even cardiovascular operations (Zhong et al. 2016; Hu et al. 2021; Masri et al. 2021). Despite numerous significant advances in solute transport in the single-phase saturated system, the characterization of aqueous tracer transport in the single-phase saturated system is fundamentally different from that when the wetting phase (e.g., water) and the non-wetting phase (e.g., air or immiscible organic liquid) coexist in the pore space (Karadimitriou et al. 2017; Hasan et al. 2020). In the unsaturated system, the transport of the aqueous tracer was significantly dependent on the complex dynamics of both the immiscible two-phase flow and the complicated wetting phase flow (Joekar-Niasar and Ataie-Ashtiani 2009; Hu et al. 2014; Karadimitriou et al. 2016). Therefore, understanding the mass transfer for the transport of the aqueous tracer in the unsaturated system was more challenging than in the single-phase saturated system.

In recent decades, the heterogeneity of porous or fractured media has been recognized as the main reason why Fick's law or the traditional Fickian transport model (i.e., the advection–dispersion equation or ADE) is unable to describe the anomalous (non-Fickian) breakthrough curves (BTCs) (Dentz et al. 2011; Jin et al. 2017b; Wang and Cardenas 2017; Dou et al. 2018a, 2019). As typical features of anomalous BTCs, the early arrival and long runout of BTCs are ubiquitous as the heterogeneous pore space creates a non-uniform flow field. The flow behavior as background information for the transport of the aqueous tracer was inseparable from the mass transfer of the aqueous tracer.

As a fundamental mass transfer process, dispersion of aqueous tracers is complex because the nonuniform flow field varies in both magnitude and direction as it meanders through the heterogeneous pore space. The heterogeneity of the medium resulted in tortuous and spatially variable flow paths. Consequently, the corresponding flow field was divided into flowing and stagnant regions, referred to as mobile and immobile regions for aqueous tracer transport. In the mobile region, mass transfer of aqueous tracer was mainly controlled by advection, whereas mass transfer in the immobile region was mainly diffusion-controlled due to poor connectivity with the main flow pathways (Illman 2014; Rolle and Kitanidis 2014; Hasan et al. 2020). The mass transfer between the mobile and immobile regions could significantly delay the transport of the aqueous tracer, which is associated with the tailing of tracer BTCs.

To this end, several remarkable models have been developed to characterize and explain the anomalous BTCs and the corresponding transport behavior (Gao et al. 2009; Neuman and Tartakovsky 2009; Dou et al. 2019; Nikan et al. 2020), such as the continuous time random walk (CTRW) (Berkowitz et al. 2006), the fractional advection–dispersion equation (FADE) (Benson 1998), the mobile-immobile model (MIM) (van Genuchten and Wierenga 1976), and the multi-rate mass transfer model (MRMT) (Haggerty and Gorelick 1995). Although fitting anomalous BTCs to different transport models is able to accurately estimate the transport parameters and quantify the characterization of non-Fickian transport behavior, the application of these models in the unsaturated system is limited and complex due to the variable pore space in immiscible two-phase flow (Bijeljic et al. 2013b; Boon et al. 2017; Ahmad et al. 2021). In particular, even with the same hydraulic gradient, the flow field can vary depending on the saturation topology. As a result, previous studies (de

Gennes 1983; Karadimitriou et al. 2016; Ramsburg 2018) have indicated that the transport behavior of aqueous tracers is sensitive to the saturation topology in the unsaturated system.

Unlike in the saturated system, the behavior of aqueous tracer transport in the unsaturated system was coupled by two basic flow processes: ① the mass transfer of the aqueous solute; ② the immiscible two-phase multiphase flow. Thus, the aqueous tracer transport in the unsaturated system was strongly influenced by the saturation of the wetting phase (e.g., water saturation) (Jiménez-Martínez et al. 2017; Ramsburg 2018; Hasan et al. 2019) and the spatiotemporal distribution of the two immiscible liquid phases (Karadimitriou et al. 2016). In general, the dispersion coefficient of the aqueous tracer is an important parameter that depends on the flow rate and water saturation. Moreover, it has been shown that the enhancement of the non-uniform flow field with decreasing water saturation leads to a dispersion process that is non-fickle to a greater extent (Bromly and Hinz 2004; Jiménez-Martínez et al. 2015). Raoof and Hassanizadeh (2013) found that the dispersivity of the aqueous tracer increases with decreasing water saturation, reaches a maximum value, and then decreases with further decreasing saturation. Recently, Hasan et al. (2020) characterized mass transfer in unsaturated porous media using fast X-ray synchrotron microtomography and showed that the dispersivity changes with water saturation. Zhuang et al. (2021) studied the effect of fluid saturation and particle size on the dispersivity of unsaturated solutes and found a non-monotonic relationship between dispersivity and water saturation. These studies could apply only to the case where both immiscible liquids are continuous in the unsaturated porous media system, but not necessarily to the case where the immiscible liquid becomes discontinuous and immobilized by capillary forces at residual saturation.

Indeed, understanding the influence of residual saturation on aqueous tracer transport is important for many industrial applications (e.g., remediation of organically contaminated groundwater) (Essaid et al. 2015; Aminnaji et al. 2019). Thus, the contaminant (e.g., a non-aqueous phase liquid, NAPL) could be immobilized at the residual NAPL saturation due to capillary forces. Consequently, the surrounding flow field was altered with both the residual NAPL saturation and the spatial distribution of residual NAPL. In previous studies (Pennell et al. 1994; Rogers and Logan 2000), increased longitudinal dispersivity of the aqueous tracer caused by residual NAPL was observed. Zhang et al. (2014) conducted solute transport experiments in sand columns contaminated with diesel and motor oil at various oil saturations. They also found that the dispersivity increased with increasing residual oil saturation, and emphasized that the immobile oil trapped in the pore space could enhance the local velocity fluctuations and increase the tortuosity of the flow paths in the aqueous phase. However, at a certain residual saturation, the spatial distribution of the residual deviated phase depends on the fluid–fluid displacement processes, which are determined by the competition between viscous and capillary forces (Lenormand et al. 1988; Dou et al. 2013; Zhao et al. 2016; Hu et al. 2017; Chen et al. 2018). The competition between these two forces leads to the basic two-phase displacement patterns, such as viscous displacement, capillary displacement, and stable displacement (Chen et al.

2017). The different biphasic displacement patterns lead to different spatial distribution of the remaining defensive phase. Therefore, it is necessary to characterize the transport of aqueous tracers through the remaining two-phase zone considering the two-phase displacement patterns.

References

Ahmad F, Rahimi A, Tsotsas E et al (2021) From micro-scale to macro-scale modeling of solute transport in drying capillary porous media. Int J Heat Mass Transf 165:120722. https://doi.org/10.1016/j.ijheatmasstransfer.2020.120722

Aminnaji M, Rabbani AJ, Niasar V et al (2019) Effects of pore-scale heterogeneity on macroscopic NAPL dissolution efficiency: a two-scale numerical simulation study. Water Resour Res 55(11):8779–8799

Anna P, Jimenez-Martinez J, Tabuteau H et al (2014) Mixing and reaction kinetics in porous media: an experimental pore scale quantification. Environ Sci Technol 48(1):508–516

Aris R (1956) On the dispersion of a solute in a fluid flowing through a tube. Proc Royal Soc London a: Mathem Phys Eng Sci 235(1200):67–77

Bagalkot N, Kumar GS (2016) Numerical modeling of two species radionuclide transport in a single fracture-matrix system with variable fracture aperture. Geosci J 20(5):627–638

Barros FP, Dentz M, Koch J et al (2012) Flow topology and scalar mixing in spatially heterogeneous flow fields. Geophys Res Lett 39(8):L08404

Bear J (2013) Dynamics of fluids in porous media. Courier Corporation, North Chelmsford

Benson DA (1998) The fractional advection-dispersion equation: development and application. University of Nevada, Reno

Berkowitz B, Cortis A, Dentz M, et al. (2006) Modeling non-Fickian transport in geological formations as a continuous time random walk. Rev Geophys 44(2). https://doi.org/10.1029/2005rg000178

Bijeljic B, Mostaghimi P, Blunt MJ (2013a) Insights into non-Fickian solute transport in carbonates. Water Resour Res 49(5):2714–2728

Bijeljic B, Raeini A, Mostaghimi P et al (2013b) Predictions of non-Fickian solute transport in different classes of porous media using direct simulation on pore-scale images. Phys Rev E 87(1):013011

Bolster D, Dentz M, Le Borgne T (2009) Solute dispersion in channels with periodically varying apertures. Phys Fluids 21(5):056601

Bolster D, Valdés-Parada FJ, LeBorgne T et al (2011) Mixing in confined stratified aquifers. J Contam Hydrol 120–121:198–212

Boon M, Bijeljic B, Krevor S (2017) Observations of the impact of rock heterogeneity on solute spreading and mixing. Water Resour Res 53(6):4624–4642

Boschan A, Ippolito I, Chertcoff R et al (2008) Geometrical and Taylor dispersion in a fracture with random obstacles: an experimental study with fluids of different rheologies. Water Resour Res 44(6):W06420

Bromly M, Hinz C (2004) Non-Fickian transport in homogeneous unsaturated repacked sand. Water Resour Res 40(7):W07402

Brown SR, Stockman HW, Reeves SJ (1995) Applicability of the reynolds equation for modeling fluid flow between rough surfaces. Geophys Res Lett 22(18):2537–2540

Brown S, Caprihan A, Hardy R (1998) Experimental observation of fluid flow channels in a single fracture. J Geophys Res: Solid Earth 103(B3):5125–5132

Brush DJ, Thomson NR (2003) Fluid flow in synthetic rough-walled fractures: navier-stokes, stokes, and local cubic law simulations. Water Resour Res 39(4):1085

Cai J, Wei W, Hu X et al (2017) Electrical conductivity models in saturated porous media: a review. Earth Sci Rev 171:419–433

Cao J, Kitanidis PK (1998) Pore-scale dilution of conservative solutes: an example. Water Resour Res 34(8):1941–1949

Cardenas MB, Slottke DT, Ketcham RA et al (2007) Navier-stokes flow and transport simulations using real fractures shows heavy tailing due to eddies. Geophys Res Lett 34(14):L14404

Cardenas MB, Slottke DT, Ketcham RA et al (2009) Effects of inertia and directionality on flow and transport in a rough asymmetric fracture. J Geophys Res 114(B6):B6204

Chen YF, Fang S, Wu DS et al (2017) Visualizing and quantifying the crossover from capillary fingering to viscous fingering in a rough fracture. Water Resour Res 53(9):7756–7772

Chen YF, Wu DS, Fang S et al (2018) Experimental study on two-phase flow in rough fracture: phase diagram and localized flow channel. Int J Heat Mass Transf 122:1298–1307

Chiogna G, Bellin A (2013) Analytical solution for reactive solute transport considering incomplete mixing within a reference elementary volume. Water Resour Res 49(5):2589–2600

Chiogna G, Hochstetler DL, Bellin A et al (2012) Mixing, entropy and reactive solute transport. Geophys Res Lett 39(20):L20405

Chiogna G, Cirpka OA, Herrera PA (2015) Helical flow and transient solute dilution in porous media. Transp Porous Media 111(3):591–603

Chiogna G, Cirpka OA, Herrera PA (2016) Helical flow and transient solute dilution in porous media. Transp Porous Media 111(3):591–603

Cirpka OA (2002) Choice of dispersion coefficients in reactive transport calculations on smoothed fields. J Contam Hydrol 58(3):261–282

Cirpka OA, Kitanidis PK (2000) Characterization of mixing and dilution in heterogeneous aquifers by means of local temporal moments. Water Resour Res 36(5):1221–1236

Cirpka OA, Valocchi AJ (2007) Two-dimensional concentration distribution for mixing-controlled bio-reactive transport in steady state. Adv Water Resour 30(6):1668–1679

de Gennes P (1983) Hydrodynamic dispersion in unsaturated porous media. J Fluid Mech 136:189–200

de Dreuzy JR, Carrera J (2016) On the validity of effective formulations for transport through heterogeneous porous media. Hydrol Earth Syst Sci 20(4):1319–1330

de Dreuzy JR, Carrera J, Dentz M et al (2012) Time evolution of mixing in heterogeneous porous media. Water Resour Res 48(6):W06511

de Simoni M, Carrera J, Sánchez-Vila X et al (2005) A procedure for the solution of multicomponent reactive transport problems. Water Resour Res 41(11):1–17

Dentz M, Carrera J (2007) Mixing and spreading in stratified flow. Phys Fluids 19(1):017107

Dentz M, de Barros FPJ (2015) Mixing-scale dependent dispersion for transport in heterogeneous flows. J Fluid Mech 777:178–195

Dentz M, Cortis A, Scher H et al (2004) Time behavior of solute transport in heterogeneous media: transition from anomalous to normal transport. Adv Water Resour 27(2):155–173

Dentz M, Le Borgne T, Englert A et al (2011) Mixing, spreading and reaction in heterogeneous media: a brief review. J Contam Hydrol 120–121:1–17

Detwiler RL, Rajaram H, Glass RJ (2000) Solute transport in variable-aperture fractures: an investigation of the relative importance of Taylor dispersion and macro-dispersion. Water Resour Res 36(7):1611–1625

Dou Z, Zhou Z, Sleep BE (2013) Influence of wettability on interfacial area during immiscible liquid invasion into a 3D self-affine rough fracture: Lattice Boltzmann simulations. Adv Water Resour 61:1–11

Dou Z, Zhou ZF, Wang JG (2016) Three-dimensional analysis of spreading and mixing of miscible compound in heterogeneous variable-aperture fracture. Water Sci Eng 9(4):293–299

Dou Z, Chen Z, Zhou Z et al (2018a) Influence of eddies on conservative solute transport through a 2D single self-affine fracture. Int J Heat Mass Transf 121:597–606

Dou Z, Zhou Z, Wang J et al (2018b) Roughness scale dependence of the relationship between tracer longitudinal dispersion and Peclet number in variable-aperture fractures. Hydrol Process 32(10):1461–1475
Dou Z, Sleep B, Zhan H et al (2019) Multiscale roughness influence on conservative solute transport in self-affine fractures. Int J Heat Mass Transf 133:606–618
Dronfield DG, Silliman SE (1993) Velocity dependence of dispersion for transport through a single fracture of variable roughness. Water Resour Res 29(10):3477–3483
Edery Y, Guadagnini A, Scher H et al (2014) Origins of anomalous transport in heterogeneous media: structural and dynamic controls. Water Resour Res 50(2):1490–1505
Engdahl NB, Ginn TR, Fogg GE (2013) Scalar dissipation rates in non-conservative transport systems. The J Contaminant Hydrol 149:46–60
Essaid HI, Bekins BA, Cozzarelli IM (2015) Organic contaminant transport and fate in the subsurface: evolution of knowledge and understanding. Water Resour Res 51(7):4861–4902
Fiori A, Janković I, Dagan G (2006) Modeling flow and transport in highly heterogeneous three-dimensional aquifers: ergodicity, gaussianity, and anomalous behavior-2. Approximate semianalytical solution. Water Resour Res 42(6): W06D13
Freyberg DL (1986) A natural gradient experiment on solute transport in a sand aquifer: 2. Spatial moments and the advection and dispersion of nonreactive tracers. Water Resour Res 22(13):2031–2046
Gao G, Zhan H, Feng S et al (2009) Comparison of alternative models for simulating anomalous solute transport in a large heterogeneous soil column. J Hydrol 377(3–4):391–404
van Genuchten MT, Wierenga P (1976) Mass transfer studies in sorbing porous media I. Analytical solutions. Soil Sci Soc America J 40(4):473–480
Gramling CM, Harvey CF, Meigs LC (2002) Reactive transport in porous media: a comparison of model prediction with laboratory visualization. Environ Sci Technol 36(11):2508–2514
Haggerty R, Gorelick SM (1995) Multiple-rate mass transfer for modeling diffusion and surface reactions in media with pore-scale heterogeneity. Water Resour Res 10:2383–2400
Haggerty R, McKenna SA, Meigs LC (2000) On the late-time behavior of tracer test breakthrough curves. Water Resour Res 36(12):3467–3479
Hasan S, Joekar-Niasar V, Karadimitriou NK et al (2019) Saturation dependence of non-Fickian transport in porous media. Water Resour Res 55(2):1153–1166
Hasan S, Niasar V, Karadimitriou NK et al (2020) Direct characterization of solute transport in unsaturated porous media using fast X-ray synchrotron microtomography. Proc Natl Acad Sci 117(38):23443
Heidari P, Li L (2014) Solute transport in low-heterogeneity sandboxes: the role of correlation length and permeability variance. Water Resour Res 50(10):8240–8264
Hou Y, Jiang J, Wu J (2018) Anomalous solute transport in cemented porous media: Pore-scale simulations. Soil Sci Soc Am J 82(1):10
Hu W, Huang N, Zhang X (2014) Impact of saturation on mass transfer rate between mobile and immobile waters in solute transport within aggregated soils. J Hydrol 519:3557–3565
Hu Y, Patmonoaji A, Xu H et al (2021) Pore-scale investigation on nonaqueous phase liquid dissolution and mass transfer in 2D and 3D porous media. Int J Heat Mass Transf 169:120901
Hu R, Wan J, Kim Y et al. (2017) Wettability effects on supercritical CO_2–brine immiscible displacement during drainage: pore-scale observation and 3D simulation. Int J Greenhouse Gas Control 60(upplement C):129–139
Illman WA (2014) Hydraulic tomography offers improved imaging of heterogeneity in fractured rocks. Groundwater 52(5):659–684
Ippolito I, Daccord G, Hinch E et al (1994) Echo tracer dispersion in model fractures with a rectangular geometry. J Contam Hydrol 16(1):87–108
Jha B, Cueto-Felgueroso L, Juanes R (2011) Quantifying mixing in viscously unstable porous media flows. Phys Rev E 84:066312

Jiménez-Martínez J, Anna PD, Tabuteau H et al (2015) Pore-scale mechanisms for the enhancement of mixing in unsaturated porous media and implications for chemical reactions. Geophys Res Lett 42(13):5316–5324

Jiménez-Martínez J, Le Borgne T, Tabuteau H et al (2017) Impact of saturation on dispersion and mixing in porous media: photobleaching pulse injection experiments and shear-enhanced mixing model. Water Resour Res 53(2):1457–1472

Jin Y, Li X, Zhao M, et al. (2017b) A mathematical model of fluid flow in tight porous media based on fractal assumptions. Int J Heat and Mass Transf 108(Part A):1078–1088

Joekar-Niasar V, Ataie-Ashtiani B (2009) Assessment of nitrate contamination in unsaturated zone of urban areas: the case study of Tehran, Iran. Environ Geol 57(8):1785–1798

Kang PK, Brown S, Juanes R (2016) Emergence of anomalous transport in stressed rough fractures. Earth Planet Sci Lett 454:46–54

Kapoor V, Kitanidis PK (1996) Concentration fluctuations and dilution in two-dimensionally periodic heterogeneous porous media. Transp Porous Media 22(1):91–119

Kapoor V, Jafvert CT, Lyn DA (1998) Experimental study of a bimolecular reaction in poiseuille flow. Water Resour Res 34(8):1997–2004

Kapoor V, Gelhar LW (1994) Transport in three-dimensionally heterogeneous aquifers: 1. Dynamics of concentration fluctuations. Water Resour Res 30(6):1775–1788

Karadimitriou NK, Joekar-Niasar V, Babaei M et al (2016) Critical role of the immobile zone in non-fickian two-phase transport: a new paradigm. Environ Sci Technol 50(8):4384–4392

Karadimitriou NK, Joekar-Niasar V, Brizuela OG (2017) Hydro-dynamic solute transport under two-phase flow conditions. Sci Rep 7(1):6624

Keller AA, Roberts PV, Blunt MJ (1999) Effect of fracture aperture variations on the dispersion of contaminants. Water Resour Res 35(1):55–63

Kitanidis PK (1994) The concept of the dilution index. Water Resour Res 30(7):2011–2026

Le Borgne T, Dentz M, Bolster D et al (2010) Non-Fickian mixing: temporal evolution of the scalar dissipation rate in heterogeneous porous media. Adv Water Resour 33(12):1468–1475

Le Borgne T, Dentz M, Villermaux E (2013) Stretching, coalescence, and mixing in porous media. Phys Rev Lett 110(20):204501

Lenormand R, Touboul E, Zarcone C (1988) Numerical models and experiments on immiscible displacements in porous media. J Fluid Mech 189(1):165–187

Luo J, Cirpka OA (2011) How well do mean breakthrough curves predict mixing-controlled reactive transport. Water Resour Res 47(2):W02520

Masri R, Puelz C, Riviere B (2021) A reduced model for solute transport in compliant blood vessels with arbitrary axial velocity profile. Int J Heat Mass Transf 176:121379

Miralles-Wilhelm F, Gelhar LW (2000) Stochastic analysis of oxygen-limited biodegradation in heterogeneous aquifers with transient microbial dynamics. J Contaminant Hydrol 42(1):69–97

Neuman SP, Tartakovsky DM (2009) Perspective on theories of non-Fickian transport in heterogeneous media. Adv Water Resour 32(5):670–680

Nikan O, Machado JAT, Golbabai A et al (2020) Numerical approach for modeling fractal mobile/immobile transport model in porous and fractured media. Int Commun Heat Mass Transfer 111:104443

Oron AP, Berkowitz B (1998) Flow in rock fractures: the local cubic law assumption reexamined. Water Resour Res 34(11):2811–2825

Pennell KD, Jin M, Abriola LM et al (1994) Surfactant enhanced remediation of soil columns contaminated by residual tetrachloroethylene. J Contam Hydrol 16(1):35–53

Pini R, Vandehey NT, Druhan J et al (2016) Quantifying solute spreading and mixing in reservoir rocks using 3-D PET imaging. J Fluid Mech 796:558–587

Qian J, Zhan H, Zhang Y et al (2015) Numerical simulation and experimental study of bimolecular reactive transport in porous media. Transp Porous Media 109(3):727–746

Raje DS, Kapoor V (2000) Experimental study of bimolecular reaction kinetics in porous media. Environ Sci Technol 34(7):1234–1239

Ramsburg KAMCA (2018) Influence of nonwetting phase saturation on dispersivity in laboratory-scale sandy porous media. Environ Eng Sci 35(10):1062–1074
Raoof A, Hassanizadeh S (2013) Saturation-dependent solute dispersivity in porous media: pore-scale processes. Water Resour Res 49(4):1943–1951
Rogers B, Logan BE (2000) Bacterial transport in NAPL-contaminated porous media. J Environ Eng 126(7):657–666
Rolle M, Kitanidis PK (2014) Effects of compound-specific dilution on transient transport and solute breakthrough: a pore-scale analysis. Adv Water Resour 71:186–199
Rolle M, Eberhardt C, Chiogna G et al (2009) Enhancement of dilution and transverse reactive mixing in porous media: experiments and model-based interpretation. J Contam Hydrol 110(3):130–142
Rolle M, Hochstetler D, Chiogna G et al (2012) Experimental investigation and pore-scale modeling interpretation of compound-specific transverse dispersion in porous media. Transp Porous Media 93(3):347–362
Rolle M, Chiogna G, Hochstetler DL et al (2013) On the importance of diffusion and compound-specific mixing for groundwater transport: an investigation from pore to field scale. J Contam Hydrol 153:51–68
Roux S, Plouraboué F, Hulin J-P (1998) Tracer dispersion in rough open cracks. Transp Porous Media 32(1):97–116
Sale TC, Zimbron JA, Dandy DS (2008) Effects of reduced contaminant loading on downgradient water quality in an idealized two-layer granular porous media. J Contam Hydrol 102(1–2):72–85
Shapiro M, Brenner H (1988) Dispersion of a chemically reactive solute in a spatially periodic model of a porous medium. Chem Eng Sci 43(3):551–571
Soltanian MR, Ritzi RW, Dai Z et al (2015) Reactive solute transport in physically and chemically heterogeneous porous media with multimodal reactive mineral facies: the Lagrangian approach. Chemosphere 122:235–244
Song X, Hong E, Seagren EA (2014) Laboratory-scale in situ bioremediation in heterogeneous porous media: biokinetics-limited scenario. J Contam Hydrol 158:78–92
Swanson RD, Binley A, Keating K et al (2015) Anomalous solute transport in saturated porous media: relating transport model parameters to electrical and nuclear magnetic resonance properties. Water Resour Res 51(2):1264–1283
Tartakovsky DM, Xiu D (2006) Stochastic analysis of transport in tubes with rough walls. J Comput Phys 217(1):248–259
Tartakovsky AM, Tartakovsky GD, Scheibe TD (2009) Effects of incomplete mixing on multicomponent reactive transport. Adv Water Resour 32(11):1674–1679
Tsang Y (1984) The effect of tortuosity on fluid flow through a single fracture. Water Resour Res 20(9):1209–1215
Tsang YW, Tsang C (1987) Channel model of flow through fractured media. Water Resour Res 23(3):467–479
Tsang Y, Tsang C, Neretnieks I et al (1988) Flow and tracer transport in fractured media: a variable aperture channel model and its properties. Water Resour Res 24(12):2049–2060
Voller VR (2015) A direct simulation demonstrating the role of spacial heterogeneity in determining anomalous diffusive transport. Water Resour Res 51(4):2119–2127
Wang L, Cardenas MB (2014) Non-fickian transport through two-dimensional rough fractures: assessment and prediction. Water Resour Res 50(2):871–884
Wang L, Cardenas MB (2015) An efficient quasi-3D particle tracking-based approach for transport through fractures with application to dynamic dispersion calculation. J Contam Hydrol 179:47–54
Wang L, Cardenas MB (2017) Transition from non-fickian to fickian longitudinal transport through 3-D rough fractures: scale-(in)sensitivity and roughness dependence. J Contam Hydrol 198:1–10
Wang L, Cardenas MB, Slottke DT et al (2015) Modification of the local cubic law of fracture flow for weak inertia, tortuosity, and roughness. Water Resour Res 51(4):2064–2080

Wang M, Chen YF, Ma GW et al (2016) Influence of surface roughness on nonlinear flow behaviors in 3D self-affine rough fractures: lattice Boltzmann simulations. Adv Water Resour 96:373–388
Xiu D, Tartakovsky DM (2006) Numerical methods for differential equations in random domains. SIAM J Sci Comput 28(3):1167–1185
Zhang J, Zheng X, Chen L et al (2014) Effect of residual oil saturation on hydrodynamic properties of porous media. J Hydrol 515:281–291
Zhao B, MacMinn CW, Juanes R (2016) Wettability control on multiphase flow in patterned microfluidics. Proc Natl Acad Sci 113(37):10251–10256
Zheng Q, Dickson S, Guo Y (2009) Influence of aperture field heterogeneity and anisotropy on dispersion regimes and dispersivity in single fractures. J Geophys Res: Solid Earth 114(B3). https://doi.org/10.1029/2007JB005161
Zhong H, El Ouni A, Lin D et al (2016) The two-phase flow IPTT method for measurement of nonwetting-wetting liquid interfacial areas at higher nonwetting saturations in natural porous media. Water Resour Res 52(7):5506–5515
Zhuang L, Raoof A, Mahmoodlu MG et al (2021) Unsaturated flow effects on solute transport in porous media. J Hydrol 598:126301
Zimmerman RW, Bodvarsson GS (1996) Hydraulic conductivity of rock fractures. Transp Porous Media 23(1):1–30
Zimmerman R, Kumar S, Bodvarsson G (1991) Lubrication theory analysis of the permeability of rough-walled fractures. Int J Rock Mech Mining Sci Geomech Abstracts 28(4):325–331
Zou L, Jing L, Cvetkovic V (2015) Roughness decomposition and nonlinear fluid flow in a single rock fracture. Int J Rock Mech Min Sci 75:102–118

Chapter 2
Concepts, Structure, and Properties of Fractured Media

Natural and artificial fractures are important channels or barriers to fluid flow in many types of media. The presence of fractures that affect flow and transport is characteristic of various types of porous media, ranging from millimeters to hundreds of kilometers. Subsurface rocks, for example, are always fractured to some degree due to tectonic movements. Fractures also exist in porous media such as soils and glaciers and in porous materials such as wood. This chapter introduces the basic concepts, structure, and properties of fractured media in geologic formations.

2.1 Basic Concepts of Fractured Media

Rock mass is a geologic formation. It has undergone many repeated geologic processes to form a particular rock composition, has been deformed and damaged, has a particular structure, and exists in a particular geologic setting. When it is used as a research object for mechanical impact, it is defined as a rock mass. Rock mass is not a simple material. It is different from rock. It is a geological body with a unique structure formed by magmatism, sedimentation and metamorphism in geological history. The composition of the rock mass consists of different minerals, and the minerals are composed of chemical components. Due to the different mineral composition and contents, there are thousands of rock masses in nature, and different rock masses have completely different physical and mechanical properties, such as granite, shale, limestone and so on. In geology, rock masses are divided into three categories: Sedimentary rock, igneous rock and metamorphic rock. In general, the internal structure of rock masses of different geneses is not exactly the same, namely their natural porosity is not exactly the same. In most cases, the gap is very small and the permeability is very low, so it is of little importance or can be ignored for the study of groundwater movement of rock masses. However, in the study of special problems such as the storage of toxic materials and nuclear waste in rocks, the permeability of

Z. Dou et al., *Mass Transfer Dynamics of Contaminants in Fractured Media*,
https://doi.org/10.1007/978-981-99-9187-7_2

natural voids in rocks cannot be ignored. The permeability of rock masses is mainly controlled by the structural plane formed by subsequent geological processes.

After their formation, natural rock masses, especially the rock masses of the Earth's crust associated with man's engineering activities, have undergone a long geological history. Most of them have been transformed by various geological functions such as tectonic movements, weathering, dissolution, and natural unloading, and will continue to change in the present and in the future. Each tectonic action leaves various traces in the rock mass: Folds, faults, fractures, etc. The physical and mechanical properties of the rock mass and its permeability depend largely on the extent and timing of tectonic activity. Multiple tectonic actions have created folds, faults and fractures in the rock mass, which have multiple overlaps and assemblages, resulting in a complex structure of the rock mass. So, it is complicated for us to study the medium properties of the rock mass groundwater. However, complexity does not mean that there is no regularity, because tectonic movements are regular, and the traces created by tectonic activity are also regular. Understanding this regularity (the nature of fractured media) is very important to the study of groundwater in fractured media, because groundwater moves mainly along the structural plane created by tectonic movement. Therefore, the study of fractured media properties is in some ways more important than the study of groundwater itself.

The smallest unit of rock mass is the boulder, that is, a boulder without significant fractures. In China, boulders are also called as structural bodies; at abroad, many scientists call boulders rock material or complete rock blocks or unit rock blocks or matrix blocks, and a few scientists call boulders rock material or rock samples. During the formation and transformation of rock masses, different structural planes have been formed. In engineering geology, Gu (1979) and Sun (1985, 1988) abstractly referred to the cracked and slightly cracked geological interfaces in the rock mass as structural planes. Broadly speaking, these are planes with a specific direction, extent, shape, and characteristics formed in the rock mass during geologic evolution. Joints, layers, and banded geologic interfaces are called structural planes, and the rock blocks intersected by the structural planes are called structural bodies, as shown in Fig. 2.1. Hydrogeology mainly studies the laws of groundwater movement in discontinuous faults and fractured structural planes.

In groundwater dynamics, a rock mass with voids is called a porous medium. Depending on the type of voids in the rock mass, porous media can be divided into porous media, fractured media, and karst-cave medium. Rock layers that contain

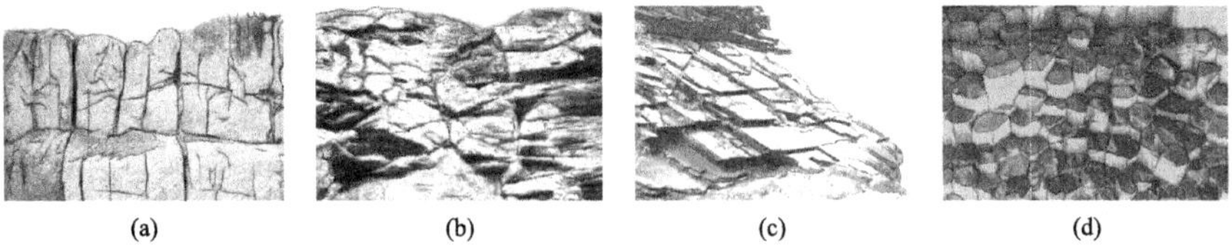

Fig. 2.1 Geologic sketch of natural fractures. **a** Blocky shape; **b** Irregular shape; **c** Plate shape; **d** Columnar shape

pore water, such as sand layers or loose sandstone, are referred to as porous media. Rock layers containing fracture water, such as quartzite and granite with developed fractures, are called fractured media. Rock masses containing karst cave water, such as limestone and dolomite with karst caves, are referred to as karst-cave media. In the broadest sense, karst cave medium can also be counted as fractured medium. Compared to porous media, the permeability of fractured media exhibits considerable anisotropy and heterogeneity due to the directionality and non-uniformity of the development and distribution of fractures (solution caves).

In fractured media there can be solid, liquid and gaseous phases. The solid phase is called the skeleton. The gaseous phase is usually air, which is mainly present in the unsaturated zone. The liquid phase is groundwater or a mixture of water and other substances or other liquids (such as oil). Groundwater can occur in many forms, such as adsorbed water, membrane water, capillary water, and gravity water. This book is mainly concerned with the movement of gravity water. The motion of fluids in fractured media is very complex and can be broadly divided into two categories: One is the motion of fluid along fractures, solution cavities, or pores in fractured media. This movement is characterized by relative dispersion and good connectivity of the fluid. The other is fluid flow along large fractures (fault zones) and pipelines, such as undergroud river in karst areas or flow along open faults. This movement is characterized by a concentration of fluid, and there are only one or more large fractures or pipelines in a large area. The flow is isolated and the groundwater flow along the fractures or pipelines is large.

Researching the causes of rock fractures can not only understand the development law of the fractured structure level to control the groundwater movement in the rock mass, but also draw conclusions about the permeability of the fractured media, which provides very important basic data and conclusions for the study of groundwater movement in the rock. The main difference between rock masses and general objects is that they are multiple fracture bodies with a certain structure, cut vertically and horizontally by the structural plane. Depending on the geological origin, fractures of rock masses can be divided into primary fractures, tectonic fractures and secondary fractures. Primary fractures refer to the structural plane formed during the formation of the rock mass, such as the primary joint formed during the cooling and contraction of the igneous rock mass, shale, gneiss, and sedimentary rock layer in metamorphic rocks. Tectonic fractures are various fracture surfaces in the rock that occurred during crustal movement after the rock had been formed, e.g., faults, joints, cleavages, etc. Secondary fractures are weathering fractures and unloading fractures caused by exogenous forces.

According to the field observations, the results of numerous mechanical tests on rock masses, and the analysis of Mohr's failure theory, the failure of rock masses during the transformation process can be divided into only two types: Shear fractures and tensile fractures, which can be divided into two categories depending on the mechanical causes of the fracture surface.

The tensile fracture surface is formed by tensile stresses. During failure, the rock mass separates on both sides of the fracture surface and shifts only in the normal direction of the fracture surface. In field geology, pinnate tensile fracture surfaces, longitudinal tensile fracture surfaces, transverse tensile fracture surfaces, cold- shrinkage fractures in igneous rocks, and dry-shrinkage fractures in clayey or argillaceous sedimentary rocks are common. The tensile fracture surface usually has a large opening, low continuity, an irregular shape, usually in the form of a dashed line or sawtooth, frequent branching and bifurcation, a non-uniform cross-section, and high roughness. The width of the fracture zone on the tensile fracture surface is large, and the width of the fracture zone changes greatly. Most tectonic rocks are loose breccias that are angular, highly variable in size, disordered, and directionless. The fracture zone is easily filled by drifts and ore veins, and sometimes magma intrudes along the fracture zone. This type of stressed fracture surface is often characterized by high water content and strong water conductivity. It is the main channel for groundwater movement.

The shear fracture surface is formed by shear stress, and the rock mass on both sides of the fracture surface has different degrees of slip displacement along the tangential direction of the fracture surface. The shear fracture surface is the most developed and complex structural surface of the rock mass. It has the common features of scratches, conjugation, displacement direction and smooth cross-section. It is one of the most important structural surfaces controlling groundwater movement in rock.

2.2 Structure of Fractured Media

The spatial distribution and natural properties of the structural plane of the rock mass can be characterized by its geometrical features. The geometric characteristics of the structural plane refer to the orientation, shape, size, spacing or density, aperture, roughness, connectivity, and filling of the structural plane. These geometric eigenvalues of the structural plane directly affect the size, direction, and law of permeability of the fractured media.

For the statistical study of the geometric parameters of the structural plane, the works before the 1980s mainly used some methods of structural geology to construct a diagram of fracture rose, a diagram of fracture poles, and a diagram of uniform density of fractures by field measurements of the elements of the occurrence of the structural plane to vividly and accurately reflect the geometric anisotropy of the fault system, group the fractures, and determine the predominant occurrence of fractures. Since the 1980s, with the rapid development of computer technology, the technology of structural plane network simulation has been used to study the probability distribution of the parameters of the structural planes of fractures. The technology of structural plane network simulation is based on the line method and the statistical window method. On the basis of a large number of fracture measurement data and through the statistical analysis of the structural plane parameters—occurrence, trace length, half trace length, aperture, spacing (density), etc. The probability distribution for each structural plane parameter is determined. On this basis, the equivalent structural plane network is determined by Monte Carlo method. The visual simulation image of the fracture system is given, and the parameters such as structural plane density, rock quality designation (RQD), permeability coefficient tensor, and fracture connectivity can be approximated. The geometrical characteristics of the structural plane in the rock mass can be expressed by the following parameters.

(1) Orientation: Orientation refers to the spatial distribution of structural planes expressed by inclination and inclination angle. The orientation of the structural plane not only determines the seepage direction of the fractured media, but also controls the failure mechanism of the rock mass. The orientation of the structural plane is generally measured with a compass or inclinometer. If the scale of the structural plane is large, or the number of structural planes to be measured is large, or the structural plane is in the region of a magnetic anomaly and the use of a magnetic needle is not possible, photogrammetry can be used. A large number of measurement results can be represented by a rose diagram or a pole isodensity diagram to group the research.
(2) Spacing: Spacing refers to the vertical distance between the same group and adjacent structural planes, which is usually the statistical average or the most frequent distance of a fracture group. The spacing of structural planes is an important index that reflects the integrity of the rock mass, the size of the rock block, and representative elementary volume (REV) of fracture medium (also called the characterization unit). Based on the measured average spacing of the structural planes, the rock mass can be described according to Table 2.1.

In engineering practice, fracture rate is often used to express the density of the fracture structure plane.

Linear fracture rate (also known as linear density or fracture frequency) (piece / m) refers to the number of fracture structural planes per unit length along the survey line direction. If the average spacing of fracture structural planes is 1 (m), then:

Surface fracture rate (also known as area density of fracture structural surface) refers to the percentage of fracture area in the unit measured area, expressed as

$$m_a = \frac{\text{summary of measured fractures area (length} \times \text{width)}}{\text{measured area of rock mass}} \times 100\%$$

Volumetric fracture rate refers to the percentage of fracture volume in the unit measured volume, expressed as

Table 2.1 Description of continuity of fracture structural plane

Description	Very narrow	Low	Medium	High	Very high
Spacing (m)	< 1.0	1.0–3.0	3.0–10.0	10.0–30.0	> 30.0

$$m_{\mathrm{v}} = \frac{\text{summary of measured fractures volume (length} \times \text{width} \times \text{aperture)}}{\text{measured volume of rock mass}} \times 100\%$$

(3) Continuity: Continuity characterizes the propagation range and extension length of the structural plane. It is one of the most important properties of fractured rock.

The continuity of the fracture structure level can be described in Table 2.1 by the continuity length measured by each structure level group.

In practice, it is very difficult to quantify continuity accurately. Because the mass of exposed rock is often less than the area or length of the continuous fracture structure plane, the actual continuity can only be estimated. Even when there are many rock exposures, the results of field investigations are subject to many difficulties and uncertainties. Therefore, the continuity of fracture structure plane is usually described by the length of the trace, i.e., the length of the discontinuity visible at the rock outcrop is measured.

(4) Roughness: Roughness refers to the roughness of the sidewall of the fracture structure plane. It is one of the most important indicators for calculating the permeability of fractures. It is generally characterized by fluctuation or fluctuation difference. Undulation is usually expressed by the undulation angle. The undulating plane of the fracture structure represents the wave difference by the distance between the crest and trough of the wave. In qualitative field description, the plane of the fracture structure can be divided into three typical sections: stepped, wavy and apartment. Each section can be divided into three types: Rough, smooth and apartment. The length of each fracture structure plane ranges generally from 1 to 10 m.
(5) Fracture aperture: Fracture aperture refers to the vertical distance between the structural plane of the fracture and the adjacent rock wall. When the rock wall is filled with clay or other fills, it includes the width of the structural plane of the fills. The fracture aperture is one of the most important geometric parameters for characterizing fracture permeability. It is usually less than 1 mm, but a large opening may occur in the following cases: First, after shear displacement of undulating cracks or fractures with high roughness; second, in tensile fractures, especially tensile fractures formed under unloading stress by valley erosion or glacial retreat; third, when the existing fractures have been transformed by erosion and dissolution. In addition, the fracture area of the rock surface resulting from manual mining increases due to blasting or discharges. A plug measurement is normally used to measure the fracture aperture. However, since it is difficult to see the small opening with the naked eye, the photographic image method can be used to measure it indirectly. The opening can be described using the terms listed in Table 2.2.
(6) Fill material: Fill material refers to the material that is filled between the adjacent walls of the fracture structure plane. The source rock of these materials is generally weak. The amount and type of fill material directly affect the overall permeability of the fractured media. Typical fills include clastic materials such as sand,

Table 2.2 Description of aperture of fracture structural plane

Description		Aperture (cm)
Closed fracture	Very close	< 0.01
	Close	0.01–0.025
	Not close	0.025–0.05
Cracked fracture	Narrow	0.05–0.10
	Medium narrow	0.10–0.50
	Wide	0.50–1.0
Open fracture	Very wide	1.0–10.0
	Extreme wide	10.0–100.0
	Cavernous	> 100.0

silt, clay, breccia, fault gouge, and mylonite, as well as chemical precipitates such as calcite, quartz, and gypsum.

The clastic material only plays the role of mechanical filling, while the calcite and quartz veins filled in the fracture play the role of cementing and healing the structural plane. Sometimes the structural plane of the fracture can be completely filled with foreign material, sometimes only partially. Regardless of whether the structural plane of the fracture is completely or only partially filled, the permeability of the fracture medium is greatly reduced owing to filler material. The permeability of all filled fracture structural planes depends mainly on the permeability of the filler materials.

(7) Number of groups: The number of groups of fracture structure planes refers to the number of fracture groups forming the transverse fracture system. According to the regularity of the distribution of fracture structure planes, they can be divided into two types: Systematic and random. The regularity of the distribution of the primary fracture structural planes and some secondary structural planes (such as structural planes and unloading fractures) is strong and belongs to the first group. They can be classified into groups according to their occurrence.
(8) Connectivity: Connectivity of fracture structure plane refers to the degree of mutual connectivity of each group of fracture systems. Connectivity includes two aspects: first, the connectivity of a single fracture structure plane, and second, the connectivity between different fracture structure planes. The connectivity of fracture structure planes mainly depends on the orientation, group numbers, size, spacing, aperture, roughness and filling of the fractures. Only connected fracture planes can be penetrated by fluid. Therefore, connectivity is of great importance when studying the permeability of fractured media. Connectivity is often described by the connectivity rate. Currently, there is no strict calculation formula, which is mostly determined by experience.

2.3 Properties of Fractured Media

2.3.1 Porosity of Rock Mass

The cavity of the rock mass includes three parts: The cavity of the rock, the fracture of the rock mass and the solution gap. The porosity is a three-dimensional space that includes pores between rock and mineral particles, pores of extrusive rocks, etc.; fractures are shallow fractures of different sizes formed in the rock mass; the solution fracture is a solution hole formed in the soluble rock mass. The porosity of the rock mass largely determines the permeability of the rock mass. In general, the more pronounced the fracture and the larger the fracture, the greater the permeability of a rock mass.

The porosity of a rock mass is usually expressed as porosity and pore ratio. Porosity is the ratio between the volume of voids in the rock sample and the volume of the whole rock sample. Since the voids of the general rock mass contain intergranular pores and fractures, the total porosity of the rock mass (n) includes the normal porosity (n_{p}) and the fracture porosity (n_{f}).

$$n = n_{\mathrm{p}} + n_{\mathrm{f}} \tag{2.3.1}$$

The size of normal porosity depends mainly on the type, composition, and structure of the rock. Most of these factors depend on the formation of the rock. The extent of fracture porosity depends on the number of fracture groups developed in the rock mass, the density and length of each fracture group, and its aperture. It is difficult to determine the fracture porosity of a rock mass. Some scholars calculate the fracture volume and fracture porosity using the shape of the hydrostatic pressure test curve and the test volume and calculating fracture porosity based on comparison of actual and theoretical propagation velocities of elastic waves. There are many fractures in the rock mass and few microfractures in the rock block. Therefore, there is few difference between the normal porosity of the rock block and its total porosity, while there is a large difference in the rock mass. The general range of variation in the total porosity of the different rock blocks is shown in Table 2.3.

Table 2.3 General variation range of total porosity of various rock blocks

Rock name	Total porosity n (%)	Rock name	Total porosity n (%)
Granite	0.2–2.0	Gneiss	0.3–2.0
Gabbro	0.1–0.4	Quartzite	0.1–2.0
Basalt	0.1–1.0	Slate	0.1–1.0
Dolerite	0.1–0.8	Marble	1.0–10.0
Rhyolite	4.0–6.0	Limestone	5.0–10.0
Andesite	10.0–15.0	Sandstone	3.0–30.0

2.3.2 Permeability of Rock Mass

In the 1980s, fractured rock masses were always treated as isotropic porous media. In construction practice, a simple water pressure test was performed in the field to determine the unit water absorption, which characterizes the average permeability of the rock with the ω-value. The value is inserted into the formula derived from the isotropic porous medium and the permeability coefficient K of the rock is converted for calculation and design.

In general, the permeability of the rock block itself is very low (see Table 2.4), and the permeability coefficient is usually 10^{-8}–10^{-7} cm/s. However, due to the development of different genetic types of fractures in the rock mass composed of rock blocks, the permeability coefficient of the fractured rock mass is much larger than that of the rock blocks, with a difference of 10^{4}–10^{7} times. The main reason for the difference in permeability coefficient between rock masses and rock blocks is the existence of fracture structure planes in the rock masses. When they cross and are connected with each other, they become a well permeable channels.

The permeability of fractured rock is closely related to the aperture, number, and spatial distribution of the fractures. In the study of groundwater movement in fractured rock strata, the permeability of the rock blocks themselves is very low and therefore can often be neglected. When the permeability of the rock block is neglected, infiltration of groundwater in fractured rock mass occurs mainly along a large number of fractures. Fractured media can be treated as continuous media or as discontinuous media. The distinction between continuous and discontinuous media is related to the scale of the research problem (as shown in Fig. 2.2). If the scale of the research problem is large and the discontinuous properties of the rock mass are insignificant, it can be considered a continuum. If the number of discontinuous structural planes in the area affected by the problem is very large and the spacing is very dense, the rock mass can be considered as a continuous medium with equivalent permeability parameters. According to the specific situation, some scholars proposed four models:

(1) Approximately impermeable rock mass with permeability coefficient close to 0, such as fresh intact bedrock.
(2) Banded permeable body with a homogeneous permeable zone of a certain width [as shown in Fig. 2.3a]. There are also anisotropic bands, as shown in Fig. 2.3b. The permeability of the clay is weak, but the fractured zones on both sides are well permeable channels.
(3) Orthotropic permeable body: If there are three main fracture groups in the bedrock that are approximately orthogonal to each other, and infiltration occurs along these fractures [as shown in Fig. 2.3c], it can be treated as an orthotropic permeable medium.
(4) For a general anisotropic permeable medium, as shown in Fig. 2.3d, it is necessary to measure the tensor of the permeability coefficient and calculate six independent components.

Table 2.4 Permeability coefficient of various rock blocks and masses

Rock samples	K (cm/s, experimental determination)	Rock types	K (cm/s, in-site measurement)
Sandstone (Cretaceous flysch)	10^{-10}–10^{-8}	Lignite bed	1.7×10^{-2}–2.4×10^{-1}
Siltstone (Cretaceous flysch)	10^{-9}–10^{-8}	Eocene limestone	10^{-4}–10^{-2}
Granite	5.0×10^{-11}–2.0×10^{-10}	Gneiss	1.2×10^{-3}–1.9×10^{-3}
Slate	7.0×10^{-11}–1.6×10^{-10}	Pegmatoid granite	6.0×10^{-4}
Breccia	4.6×10^{-10}	Marble of Jinping I hydropower station	8.0×10^{-6}–3.2×10^{-3}
Calcite	7.0×10^{-10}–9.3×10^{-8}	Basalt of Baihetan Hydropower station	2.2×10^{-6}–4.3×10^{-5}
Limestone	7.0×10^{-10}–1.2×10^{-7}	Mudstone in chaohu power plant, Anhui Province	3.4×10^{-6}–1.1×10^{-4}
Dolomite	4.6×10^{-9}–1.2×10^{-8}	Sandstone in micro-weathering zone of Guanyinyan Hydropower Station	3.8×10^{-5}–5.9×10^{-4}
Sandstone	1.6×10^{-7}–1.2×10^{-5}	Siltstone of hongping pumped storage power station in Jiangxi Province	8.1×10^{-6}–3.7×10^{-4}
Hard mudstone	6.0×10^{-7}–2.0×10^{-6}	Quartz sandstone of Liyang pumped storage power station	2.2×10^{-5}–1.6×10^{-4}
Black schist (with cracks)	1.0×10^{-4}–3.0×10^{-4}	Basalt of Xiluodu hydropower station	7.5×10^{-7}–4.3×10^{-4}
Bradfort sandstone	6.0×10^{-7}–2.2×10^{-5}	Granite of Three gorges hydropower station	2.6×10^{-7}–1.2×10^{-3}
Glenrose sandstone	1.3×10^{-4}–1.5×10^{-3}		
Altered granite	1.5×10^{-5}–6.0×10^{-1}		

The first model is not the main research object of seepage problem in fractured rock mass from the perspective of engineering. The second model is a single fracture hydraulic problem. As early as the early 1950s, scholars began a large number of experimental studies. For the third and fourth models, when the number of fractures in each group is rare, the discrete network model or dual medium model can be used to study the groundwater movement in rock mass; when there are many fractures in each group, the fractured rock mass is generally assumed to be a continuous medium. It is considered that the flow velocity of water in the fracture becomes the hypothetical

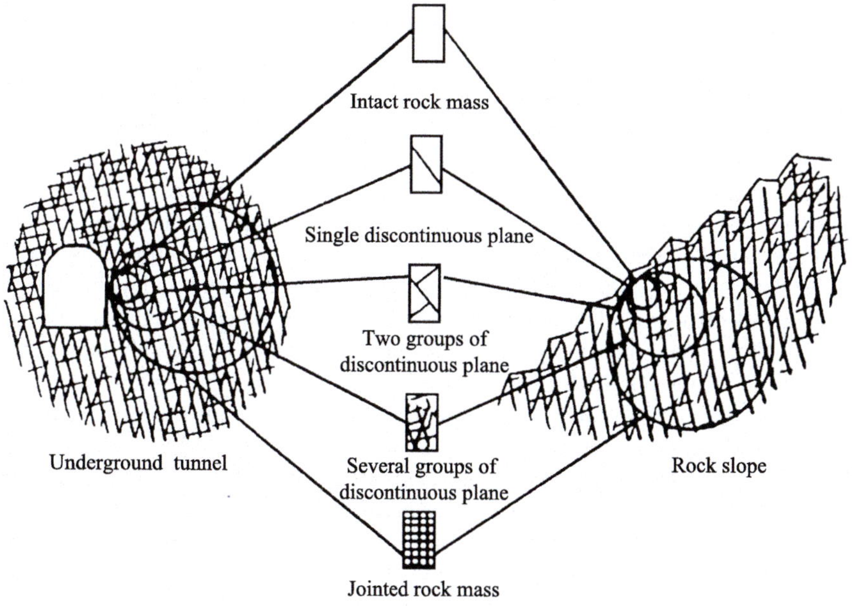

Fig. 2.2 Influence of relative size on permeability of rock mass

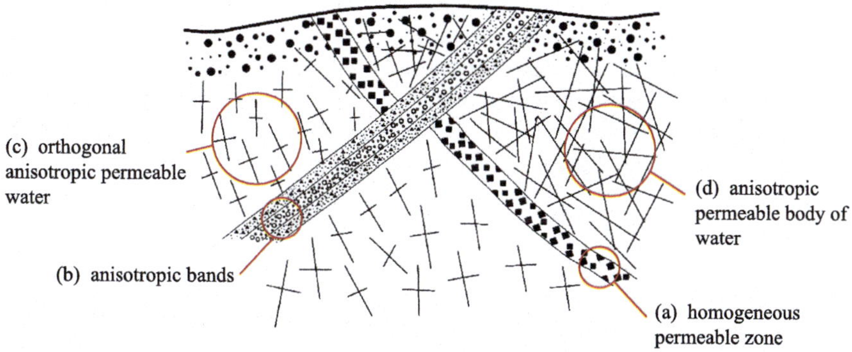

Fig. 2.3 Groundwater seepage model of rock mass

flow velocity that continuously fills the whole rock mass. The seepage problem is analyzed according to the continuous medium model.

2.3.3 Permeability of Geologic Formations

As an objective geologic formation, the permeability of the actual fractured rock mass is also controlled by weathering, unloading, landform, and other factors. Numerous analyses and studies of field data show that the permeability of the rock mass gradually decreases from the surface to depth as weathering and unloading weaken. For example, 96 borehole data of water pressure tests in the basalt fracture in the Xiluodu hydropower plant dam were analyzed and counted section by section. The results show that in the unloading zone of the valley slopes on both banks and within a depth of about 30–60 m around the valley, the rock mass is dominated by high permeability (10–100 Lu), some are highly permeable (> 100 Lu) and moderately permeable-slightly permeable (1–10 Lu) (as shown in Figs. 2.4 and 2.5). Among them, the permeable rock mass with stronger permeability accounts for 69.03%, while the medium and micro-permeable media accounts for only 30.97%. The weakly weathered rock mass in the relief zone with a depth of about 60–100 m around the valley is mainly moderately permeable-slightly permeable, and the interlayer, interlayer dislocation zone, influence zone and crack-tight zone are strongly permeable. The slightly weathered-fresh rock mass within 60–100 m is mainly slightly permeable with Lu value less than 3 Lu, except that some sections are strongly permeable (as shown in Figs. 2.6 and 2.7).

Another example is the analysis of the exploration data of granite in the dam area of Jinshuitan hydropower station and volcanic tuff in the dam area of Shitang hydropower station in Zhejiang Province. Here, it can be seen that the water absorption of the drilled rock decreases significantly with increasing depth (as shown in Fig. 2.8). Thus, the vertical structural change of rock mass in a natural geological body causes the movement of fractured groundwater to be heterogeneous in the vertical direction and to have a certain zonation.

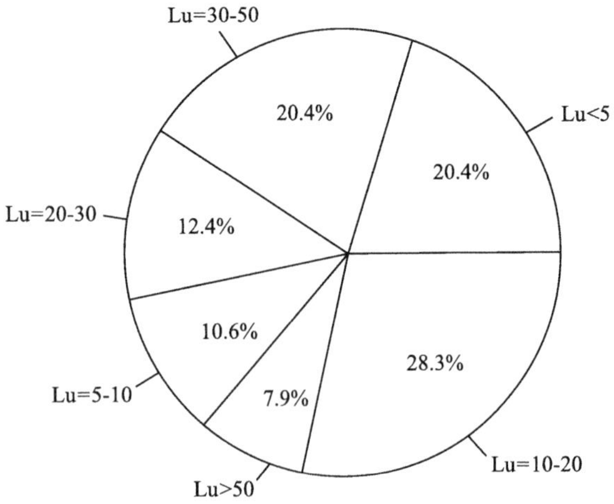

Fig. 2.4 Statistical pie chart of basalt permeability around Xiluodu Valley

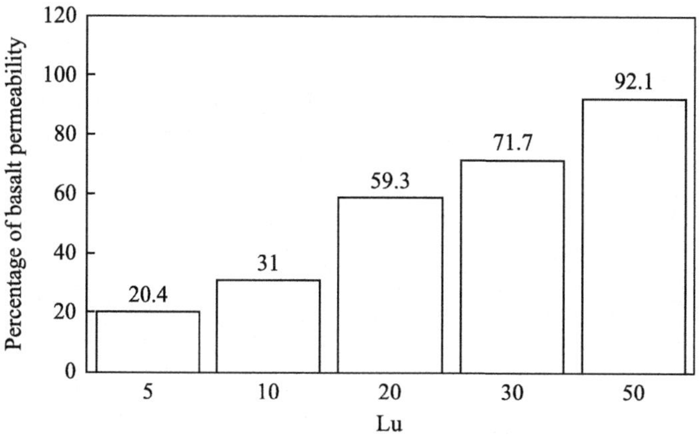

Fig. 2.5 Percentage of basalt permeability around Xiluodu Valley less than a Lu value test section

Fig. 2.6 Statistical pie chart of basalt permeability below the weak weathering zone around Xiluodu Valley

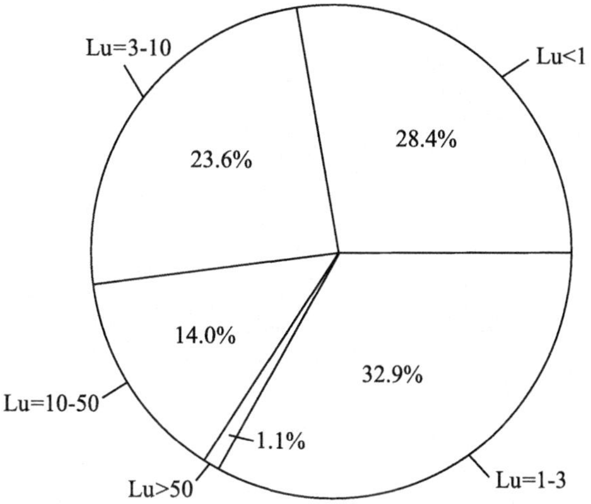

In addition, geomorphic units also have a great influence on the permeability of geologic formations. For example, the permeability of slightly weathered and fresh rocks in the dam area of the Three Gorges Project varies greatly in different geomorphic units. The general trend is that the permeability of the rocks gradually increases from the mountains on both banks toward the riverbed (as shown in Table 2.5). The permeability of the slightly weathered and fresh rock masses in the upland section on both banks is very low. The test section with very low permeability accounts for 86% to 88% of the total length of the test section. Because of the thick weathering crust, the slightly weathered and fresh rock masses are only slightly affected by unloading and groundwater flows mainly in the rock masses above the slightly weathered zone. In the riverbed zone, there is no weathered crust or the weathered crust is very thin. The slightly weathered and fresh rock masses are obviously affected by unloading

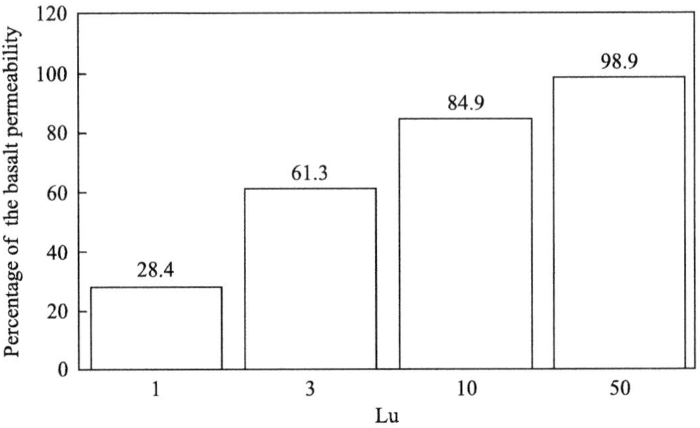

Fig. 2.7 Percentage of the basalt permeability below the weak weathering zone around Xiluodu Valley that less than a Lu value test section

Fig. 2.8 Variation curve of unit water absorption of rock mass with depth

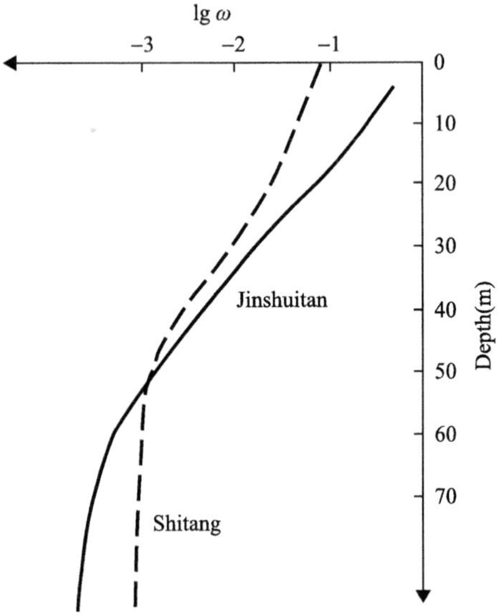

and are in the river channel of long-term water flow. The groundwater circulation changes greatly, and the permeability is significantly increased. The test section with very low permeability only accounts for 39–49% of the total test section. The weathering, unloading and hydrogeological conditions of the floodplain and the Houhe River are similar to those of the riverbed, so the permeability of the rock mass is also large.

Table 2.5 Rock permeability statistics table

Weathering strip	Landform units	Percentage of permeability classification in total length of test section (%)				
		Extremely weak permeability Lu < 1	Weak permeability Lu = 1–5	Medium permeability Lu = 5–10	Relatively Severe permeability Lu = 10–100	Severe permeability Lu = 100–1000
Slightly weathering zone	River bed	39.0	20.1	15.4	23.0	2.5
	Flood plain	68.8	16.7	6.5	6.8	0.2
	First grade terrace	86.1	9.7	1.7	2.1	0.4
	Mountain	87.2	8.6	2.0	1. 9	0.3
	Total average	79.4	12.0	3.8	4.4	0.4
Fresh rocks	River bed	48.8	41.7	7.1	2.4	0
	Flood plain	78.4	17.6	1.7	2.3	0
	First grade terrace	88.0	9.6	1.3	1.1	0
	Mountain	94.5	5.5	0	0	0
	Total average	83.5	13.5	1.7	1.3	0

2.4 Characterization and Reconstruction of Fractured Media

2.4.1 2D Self-affine Fracture Generation

Previous studies (Mandelbrot 1983) on the morphology of natural fracture walls indicated that the walls of natural fractures could be characterized as statistical self-affine distributions. We briefly review here the mathematical characterization of the self-affine rough fracture wall. We consider a two-dimensional single fracture, whose height is defined by a single-value function $Z(x)$ and the statistical self-affine property of the height can be expressed as

$$\lambda^H Z(x) = Z(\lambda x) \tag{2.4.1}$$

where H indicates the magnitude of the roughness or the so-called Hurst exponent varying from 0 to 1 and λ is a scaling factor. $Z(x)$ can be thought of as a function of an independent spatial or temporal variable x. For a self-affine fracture wall, the stationary increment $[Z(x + h_x) - Z(x)]$ over the distance h_x follows a Gaussian distribution with mean zero. Thus, for the arbitrary λ, the mean and variance of the increments can be expressed as

$$\langle Z(x + \lambda h_x) - Z(x) \rangle = 0 \quad (2.4.2)$$

$$\sigma^2(\lambda) = \lambda^{2H} \sigma^2 \quad (2.4.3)$$

where $\langle \cdot \rangle$ represents the mathematical expectation. For the different distances, the variance $\sigma^2(\lambda)$ is defined as a function of λ

$$\sigma^2(\lambda) = \left\langle [Z(x + \lambda) - Z(x)]^2 \right\rangle \quad (2.4.4)$$

$$\sigma^2(\lambda h_x) = \left\langle [Z(x + \lambda h_x) - Z(x)]^2 \right\rangle \quad (2.4.5)$$

Then, depending on Eq. (2.4.3), we have

$$\sigma_{\lambda h_x}^2 = \lambda^{2H} \sigma_{h_x}^2 \quad (2.4.6)$$

$$\sigma_{\lambda h_x} = \lambda^H \sigma_{h_x} \quad (2.4.7)$$

where $\sigma_{\lambda h_x}^2$ and $\sigma_{h_x}^2$ represent the variances of increments with different distances λh_x and λ, respectively.

Based on the self-affine scaling law given by Eqs. (2.4.1-2.4.7), a number of algorithms (e.g., the successive random additions, the randomization of the Weierstrass-Mandelbrot function, and the Fourier transformation) have been developed for synthetic self-affine fracture generation. In the present study, the successive random addition algorithm (Dou et al. 2013) was used to generate the synthetic self-affine fracture wall. It should be noted that to generate the self-affine fracture wall, the desired Hurst exponent must be selected. For this, we measured the rough surface morphology of the bottom fracture wall of a single-fracture dolomite rock block (of 280 × 210 × 70 mm in size) using a 3D stereo-topometric measurement system (ATOS II from GOM mbH, Germany). Preparation of this single-fracture dolomite rock block has been described (Mondal and Sleep 2012). A 3D model of the fracture wall surface was generated by ATOS II using non-contact optical scanning technique (Tatone and Grasselli 2009). A raw data set of fracture surface heights from the dolomite rock fracture sample was obtained with a spatial resolution of ~ 250 μm (see Fig. 2.9).

From variogram analysis (Develi and Babadagli 1998), we evaluated the self-affinity of dolomite rock surface and examined the distribution of Hurst exponent

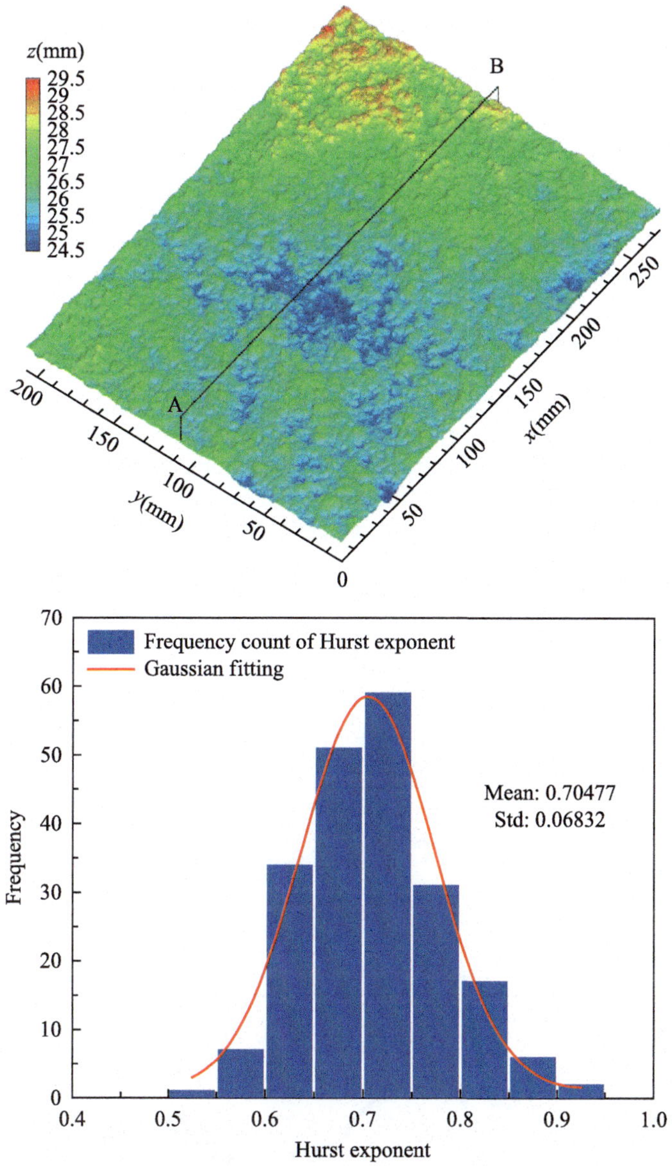

Fig. 2.9 Dolomite rock fracture surface 3D profile and distribution of Hurst exponent for the different 2D profiles along the longitudinal direction

values over 200 profiles along the longitudinal direction (for example, A-B profile in Fig. 2.9). The calculated Hurst exponent values were between 0.55 and 0.91, where Hurst exponent between 0.6 and 0.8 covers ~ 90% of profiles. Thus, due to the computational cost, we selected three different Hurst exponents, $H = 0.6$, $H = 0.7$, and $H = 0.8$.

Once the desired Hurst exponent was selected, the self-affine fracture wall was generated by the successive random addition algorithm. Figure 2.10 shows the generated self-affine fracture walls with $H = 0.8$, $H = 0.7$, and $H = 0.6$, respectively. It can be seen from Fig. 2.10 that the larger Hurst exponent leads to a higher spatial correlation and a smoother wall.

In self-affine fractures, the aperture distribution can have a significant influence on the spreading and mixing processes. The reconstruction of the aperture field from a pair of generated fracture walls is dependent upon the way in which the walls are oriented relative to each other. Note that the Hurst exponent could be different for the top and bottom fracture walls. However, our aim here is not to perform an exhaustive investigation for all possibilities. We assume that the reconstructed fracture is two-dimensional, uniform mineral component, there is no contact area (no zero-aperture region), and the Hurst exponent for both of the top and bottom fracture walls is the same. We consider two possibilities to reconstruct the rough

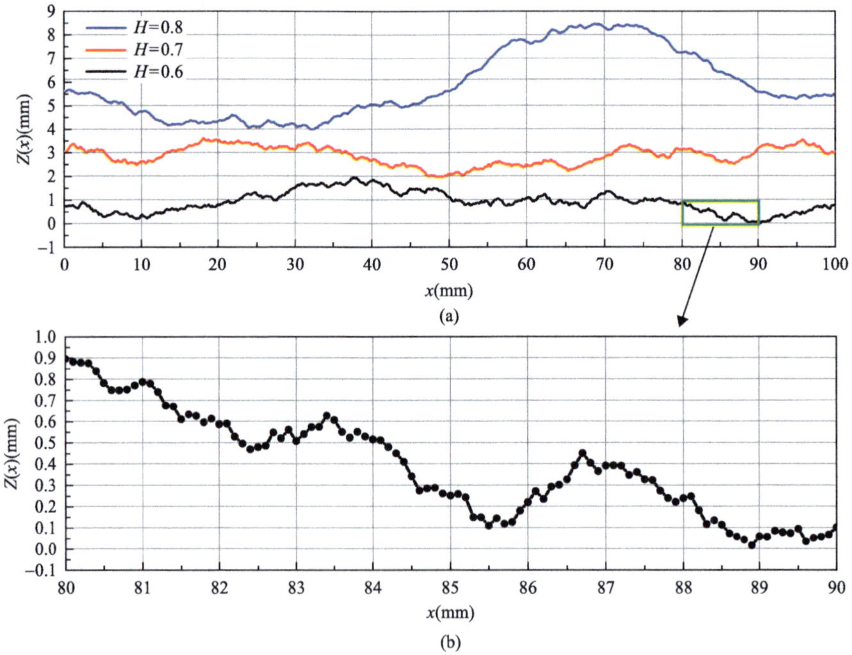

Fig. 2.10 Self-affine fracture walls with the different Hurst exponents **a** The self-affine fracture walls with $H = 0.6$, $H = 0.7$, and $H = 0.8$, respectively; **b** The zoom-in self-affine fracture wall with $H = 0.6$ between $x = 80$ mm and $x = 90$ mm

fracture aperture field from the specific self-affine fracture walls. First, we simulated a constant-aperture rough fracture where the top fracture wall was a replica of the bottom fracture wall translated a distance $\overline{b}$ normal to the mean plane [see Fig. 2.11a]. In this case, the fracture walls are rough and self-affine, but the local aperture $b(x)$ is constant and equal to $\overline{b}$. Alternatively, we studied the variable-aperture rough fracture where the top fracture wall was a replica of the bottom wall that was sheared along the horizontal direction by a displacement d_0 and then translated a distance $\overline{b}$ normal to the mean plane [see Fig. 2.11b]. Obviously, the two walls of the reconstructed fracture with shear displacement d_0 do not overlap and the local aperture is a function of the horizontal location x. Based on the self-affine scaling law, Wang et al. (1988) developed the shear displacement model to obtain the aperture field with Gaussian distribution. The local aperture $b(x)$ is given by

$$S_2(x) = S_1(x + d_0) + \overline{b} \tag{2.4.8}$$

$$b(x) = \begin{cases} S_2(x) - S_1(x), & \text{if } S_2(x) > S_1(x) \\ 0, & \text{otherwise} \end{cases} \tag{2.4.9}$$

where $S_1(x)$ and $S_2(x)$ are the top and bottom fracture walls, respectively. Figure 2.11c shows that when the Hurst exponent of the self-affine fracture wall is $H = 0.7$, the aperture field of the variable-aperture rough fracture follows a Gaussian distribution with the mean aperture $\overline{b} = 0.5$ mm and the standard deviation of the aperture $\sigma_b = 0.17$ mm.

2.4.2 3D Sheared Fractures with the Shear Displacement

There are three steps to construct a 3D sheared fracture with a certain shear displacement. First, a 3D self-affine rough fracture surface was generated. Then the generated 3D fracture surface was considered as the lower fracture surface and the upper fracture surface was translated a vertical distance without any shear displacement. In this step, although both the lower and upper surfaces were rough, the aperture in the constructed 3D fracture was constant. Lastly, the shear process was simulated by moving the upper surface to a certain displacement and keeping the lower surface fixed. A more detailed application for constructing the 3D sheared fracture can be found in previous work (Liu et al. 2020).

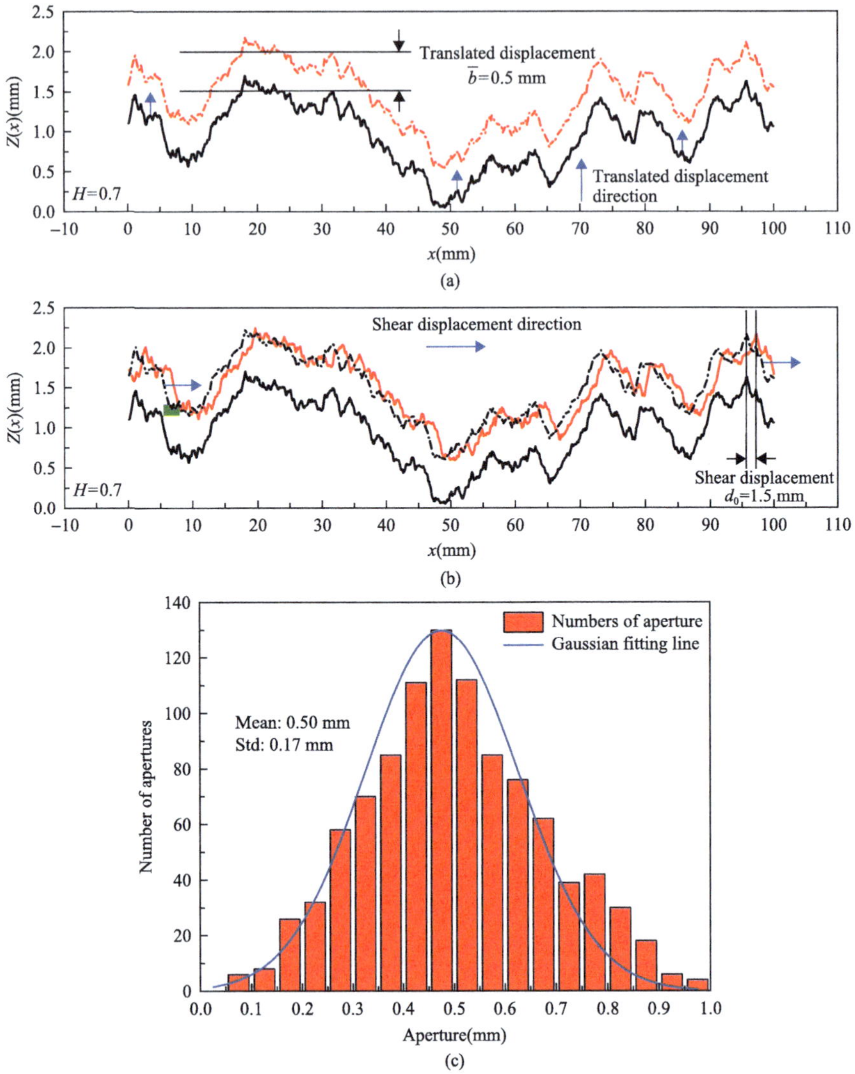

Fig. 2.11 Reconstruction of aperture field in the self-affine rough fracture with $H = 0.7$. **a** Reconstruction of constant-aperture rough fracture; **b** Reconstruction of variable-aperture rough fracture; **c** Gaussian distribution of aperture field for the variable-aperture rough fracture

Previous studies (Jin et al. 2017a, 2019; Dou et al. 2019; Liu et al. 2020) showed that the aperture in the rough fracture naturally follows a self-affine fractal distribution, implying that the roughness of the rough fracture surface exhibits a repeating pattern at different scales. A 3D self-affine fracture was generated with the technique of successive random additions (Voss 1988). As an efficient and fast algorithm, the fractional Brownian motion was employed based on the technique of successive random additions. It should be noted that as a key characteristic parameter, the Hurst exponent, H, is related to the spatial correlation and the roughness of fracture surface. Here, a 3D self-affine rough fracture surface with the size of 14 mm length and 10 mm width was generated by selecting $H = 0.5$, since the H varies between 0.45–0.85 for a natural rock fractures (Babadagli et al. 2015).

Once the 3D self-affine rough fracture surface was generated, the generated rough fracture surface was constructed to a 3D rough fracture with a constant aperture. The shear process was implemented with three different shear displacements (e.g., $U_s = 2.2$ mm, 2.6 mm and 3.0 mm). It should be mentioned that to ensure the consistency of the constructed 3D sheared fracture, a smaller zone with the size of 10 mm length and 10 mm width in the constructed 3D rough fracture was selected since the shear displacement followed the longitude direction of the generated rough fracture surface. Figure 2.12 showed the constructed sheared fracture with three different shear displacements. From the Fig. 2.12, it can be seen that as the shear displacement increases, not only the void spaces in the sheared fracture changes but also the contact area increases. The aperture distribution of the sheared fracture with three different shear displacements was shown in Fig. 2.12. The average apertures are 0.29 mm, 0.48 mm, and 0.67 mm in the sheared fracture with the shear displacement $U_s = 2.2$ mm, 2.6 mm and 3.0 mm, respectively.

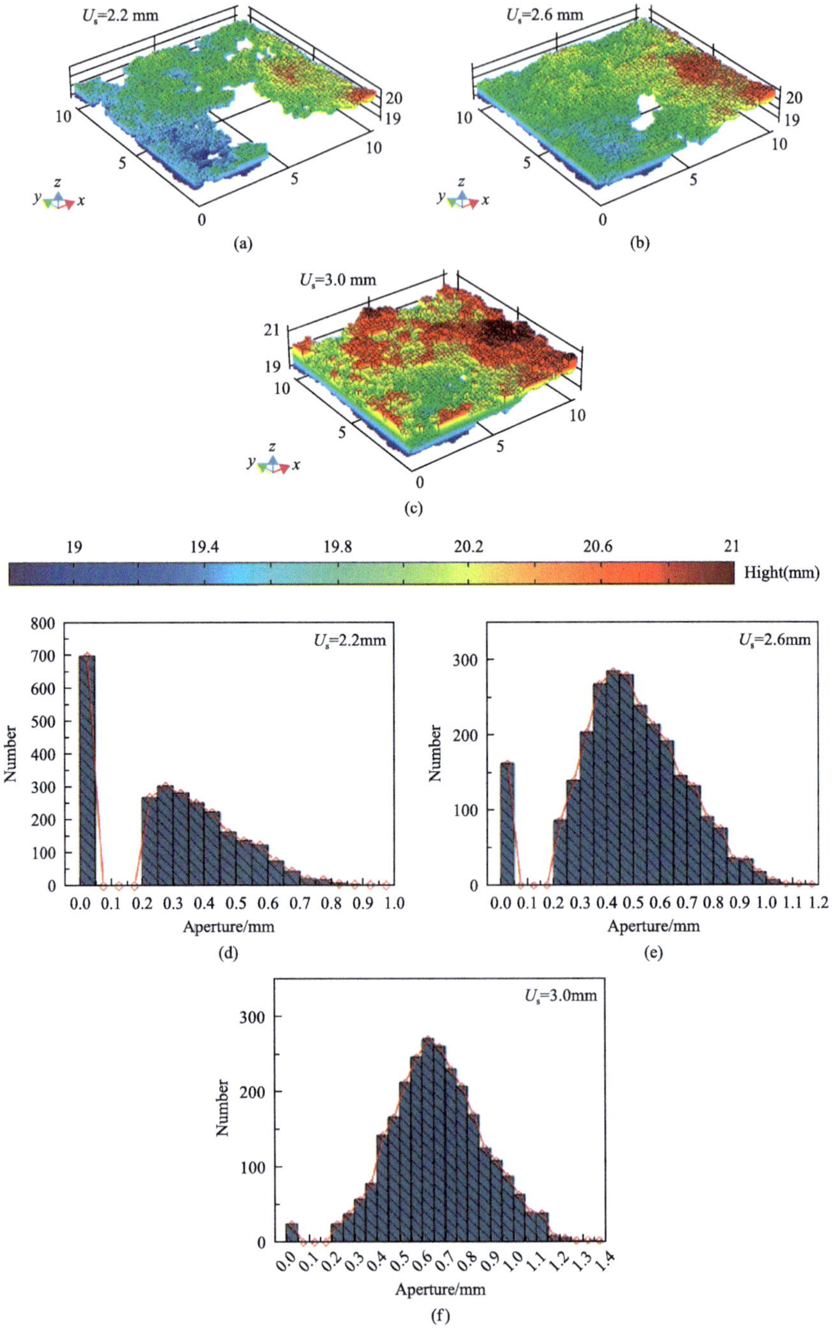

Fig. 2.12 The constructed sheared fractures with three different shear displacements and the corresponding aperture distribution

References

Babadagli T, Ren X, Develi K (2015) Effects of fractal surface roughness and lithology on single and multiphase flow in a single fracture: an experimental investigation. Int J Multiph Flow 68:40–58

Develi K, Babadagli T (1998) Quantification of natural fracture surfaces using fractal geometry. Math Geol 30(8):971–998

Dou Z, Zhou Z, Sleep BE (2013) Influence of wettability on interfacial area during immiscible liquid invasion into a 3D self-affine rough fracture: Lattice Boltzmann simulations. Adv Water Resour 61:1–11

Dou Z, Sleep B, Zhan H et al (2019) Multiscale roughness influence on conservative solute transport in self-affine fractures. Int J Heat Mass Transf 133:606–618

Gu DZ, Huang DC (1979) Classification of rock mass structure and determination of its quality coefficient. Hydrogeol Eng Geol2: 8–13

Jin Y, Dong J, Zhang X et al (2017) Scale and size effects on fluid flow through self-affine rough fractures. Int J Heat Mass Transf 105:443–451

Jin Y, Liu X, Song H et al (2019) General fractal topography: an open mathematical framework to characterize and model mono-scale-invariances. Nonlinear Dyn 96(4):2413–2436

Liu R, He M, Huang N et al (2020) Three-dimensional double-rough-walled modeling of fluid flow through self-affine shear fractures. J Rock Mech Geotech Eng 12(1):41–49

Mandelbrot BB (1983) The fractal geometry of nature. Am J Phys 51(3):286

Mondal PK, Sleep BE (2012) Colloid transport in dolomite rock fractures: effects of fracture characteristics, specific discharge, and ionic strength. Environ Sci Technol 46(18):9987–9994

Sun GZ, Huang YF (1988) A case study of wall rock panel cracking of underground cavern with high side wall and its mechanical analysis. Chinese J Rock Mech Eng 1:15–24

Sun GZ, Zhang WB (1985) Common rock mass structure: slab fracture structure and its mechanical model. Chinese J Geol 3:275–282

Tatone BS, Grasselli G (2009) A method to evaluate the three-dimensional roughness of fracture surfaces in brittle geomaterials. Rev Sci Instrum 80(12):125110

Voss RF (1988) Fractals in nature: from characterization to simulation, the science of fractal images. Springer, New York

Wang JSY, Narasimhan TN, Scholz CH (1988) Aperture correlation of a fractal fracture. J Geophys Res: Solid Earth 93(B3):2216–2224

Chapter 3
Basic Law of Fluid Flow in Fractured Media

In the last decade, the study of flow in fractured porous media has proliferated, particularly in the context of subsurface processes, but also in materials science and biological applications, as interconnected fractures can completely dominate flow patterns. Since fractures are approximately planar discontinuities with an extreme size-to-width ratio, they pose a challenge for mathematical and numerical modeling of macroscale flows based on averaging. Therefore, various and fundamentally different modeling approaches have been developed in the last decades. This chapter reviews common conceptual models and discretization approaches for flow in fractured porous media, focusing on the dominant effects of fractures on flow processes. This chapter discusses the close connection between physical and mathematical modeling and simulation approaches. Extensions and research challenges are related to flow transport and multiphase flow.

3.1 Basic Concepts of Fluid Flow in Fractured Media

Mechanics is the oldest physical science dealing with both bodies at rest and bodies in motion under the influence of forces. The branch of mechanics that deals with bodies at rest is called statics, while the branch that deals with bodies in motion under the influence of forces is called dynamics. The subcategory fluid mechanics is defined as the science that deals with the behavior of fluids at rest (fluid statics) or in motion (fluid dynamics) and the interaction of fluids with solids or other fluids at the interfaces.

Fluid mechanics is also divided into several categories. The study of the motion of fluids that can be considered incompressible (such as liquids, especially water, and gasses at low velocities) is usually referred to as hydrodynamics. A subcategory of hydrodynamics is hydraulics, which deals with the flow of fluids in pipes and open channels. Gas dynamics deals with the flow of fluids that undergo significant

Z. Dou et al., *Mass Transfer Dynamics of Contaminants in Fractured Media*,
https://doi.org/10.1007/978-981-99-9187-7_3

density changes, such as the flow of gasses through nozzles at high velocities. The aerodynamics category deals with the flow of gasses (especially air) over bodies such as airplanes, rockets, and automobiles at high or low velocities. Some other specialized categories such as meteorology, oceanography, and hydrology deal with naturally occurring flows.

Previously, we defined fluid mechanics as the science that deals with the behavior of fluids at rest or in motion and the interaction of fluids with solids or other fluids at the boundaries. There are a variety of flow problems that arise in practice, and it is usually useful to classify them on the basis of some common characteristics so that they can be studied in groups. There are many ways to classify flow problems, and we present here some general categories. In this chapter, we introduce the basic concepts commonly used in flow analysis and discuss the linear flow law, the nonlinear flow law, and multiphase flow in fractured media.

3.1.1 Viscous Versus Inviscid Regions of Flow

When two layers of fluid move relative to each other, a frictional force is created between them, and the slower layer tries to slow down the faster one. This internal flow resistance is quantified by the fluid property viscosity, which is a measure of the internal stickiness of the fluid. Viscosity is caused by cohesive forces between molecules in liquids and by molecular collisions in gases. There is no fluid with zero viscosity, and therefore all fluid flows involve viscous effects to some degree. Flows in which the frictional effects are significant are called viscous flows. However, in many flows of practical interest, there are regions (typically not near solid surfaces) where viscous forces are negligible compared to inertial or pressure forces. Neglecting the viscous terms in such non-viscous flow regions greatly simplifies the analysis without significantly affecting accuracy.

3.1.2 Laminar Versus Turbulent Flow

Some flows are smooth and orderly, while others are more chaotic. The very orderly fluid motion characterized by smooth fluid layers is called laminar. The word laminar comes from the motion of adjacent fluid particles in "lamellae". The flow of high viscosity fluids, such as oils, at low velocities is typically laminar. The highly disordered fluid motion that typically occurs at high velocities and is characterized by velocity fluctuations is called turbulent. The flow of low viscosity fluids, such as air, at high velocities is typically turbulent. Flows that are alternately laminar and turbulent are called transitional flows. Experiments conducted by Osborne Reynolds in the 1880s led to the establishment of the dimensionless Reynolds number, *Re*, as a key parameter for determining flow behavior in pipes.

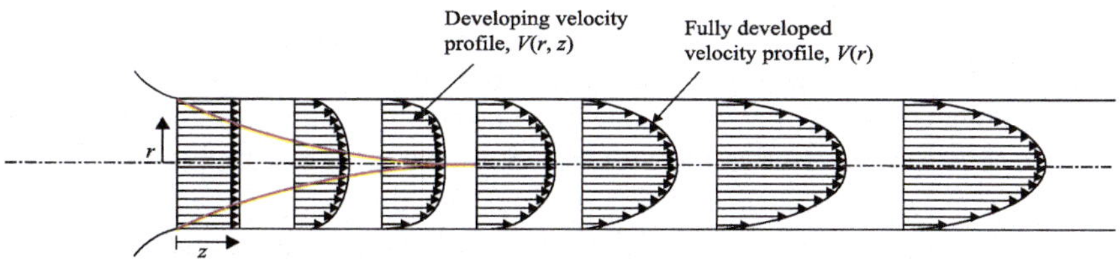

Fig. 3.1 The development of the velocity profile in a circular pipe $V = V(r, z)$ and thus the flow is two-dimensional in the entrance region, and becomes one-dimensional downstream when the velocity profile fully develops and remains unchanged in the flow direction, $V = V(r)$

3.1.3 One, Two, and Three-Dimensional Flows

A flow field is best characterized by its velocity distribution. Therefore, a flow is said to be one-, two-, or three-dimensional if the flow velocity varies in one, two, or three primary dimensions, respectively. A typical fluid flow has a three-dimensional geometry and the velocity can vary in all three dimensions, making the flow three-dimensional [$V\ (x, y, z)$ in rectangular or $V\ (r, \theta, z)$ in cylindrical coordinates]. However, the variations in velocity in certain directions may be small compared to the variations in other directions and can be ignored with negligible error. In such cases, the flow can be conveniently modeled as one- or two-dimensional, which is easier to analyze.

Consider the uniform flow of a fluid entering a circular pipe from a large tank. The velocity of the fluid is zero everywhere on the pipe surface because there is no slip. In the entrance region of the pipe, the flow is two-dimensional since the velocity changes in both the r- and z-directions, but not in the θ-direction. The velocity profile fully evolves and remains unchanged after a certain distance from the inlet (about 10 pipe diameters for turbulent flow and typically farther than for laminar pipe flow, as in Fig. 3.1), and the flow in this region is called fully evolved. The fully developed flow in a circular pipe is one-dimensional because the velocity changes in the radial r-direction, but not in the θ-angle direction or the axial z-direction, as shown in Fig. 3.1. That is, the velocity profile is the same at each axial z-position and symmetrical about the tube axis.

3.2 Linear Flow Law

3.2.1 Darcy's Law

Darcy's law describes the principle governing the motion of fluids in a given substance. Darcy's Law equation describes the ability of a fluid to flow through a porous medium such as a rock. The law is based on the fact that the flow between two points is directly proportional to the pressure differences between the points, the distance, and the permeability of the flow in the rock between the points. The measurement of permeability is called permeability.

One application of Darcy's law is the flow of water through an aquifer. Darcy's law with the equation of conservation of mass is equivalent to the equation of groundwater flow and is one of the fundamental relationships of hydrogeology. Darcy's law is also used to describe the flow of oil, gas, and water through petroleum reservoirs.

Fluid flow within the rock is determined by the permeability of the rock. Permeability must be determined in both horizontal and vertical directions. For example, shale rock consists of improbabilities that are lower vertically. This means that fluid cannot flow as easily up and down through the shale bed, but can flow more easily from one side to the other.

Darcy's law can be applied to many situations, but they are not consistent with these assumptions.

(1) Unsaturated and saturated flow.
(2) Flow in fractured rock and granular media.
(3) Unstable flow and steady-state flow.
(4) Flow in aquitards and aquifers.
(5) Flow in homogeneous and heterogeneous systems.

Darcy's law is critical to understanding many areas of geology, especially hydrogeology. Before discussing the law and its significance, we should look at how it was developed. Darcy's Law is named after Henry Darcy, a nineteenth century French engineer who developed an underground penstock system to supply water to the city of Dijon. The system, which also supplied water to Dijon's famous fountains, revolutionized municipal water and wastewater systems. The system required no pumps and was powered solely by gravity. During the development of the new system, Darcy conducted a series of experiments in which he attempted to move water by gravity alone (Fig. 3.2).

During the experiments he collected data on the length of the flume(L), the cross sectional area of the flume (A), the height difference (h_1–h_2), and the flux of water coming out Q both with and without granular (filtering) material in the flume. He then plotted the flux measurement normalized to the area against the ratio of the difference in elevation to the length of the flume [$(h_1$–$h_2)/L$]. He found that there is a linear relationship, and the slope depends on the granules he used in the flume:

$$\frac{Q}{A} \propto \frac{h_1 - h_2}{L} \tag{3.2.1}$$

This can be rewritten as follows:

$$Q = -KA\frac{\mathrm{d}h}{\mathrm{d}x} \tag{3.2.2}$$

where K is the permeability coefficient and $\mathrm{d}h/\mathrm{d}x$ is the hydraulic gradient. Q is the total discharge rate (m^3/s). The negative sign is due to the fluid flowing down the hydraulic gradient from higher values to lower values (negative).

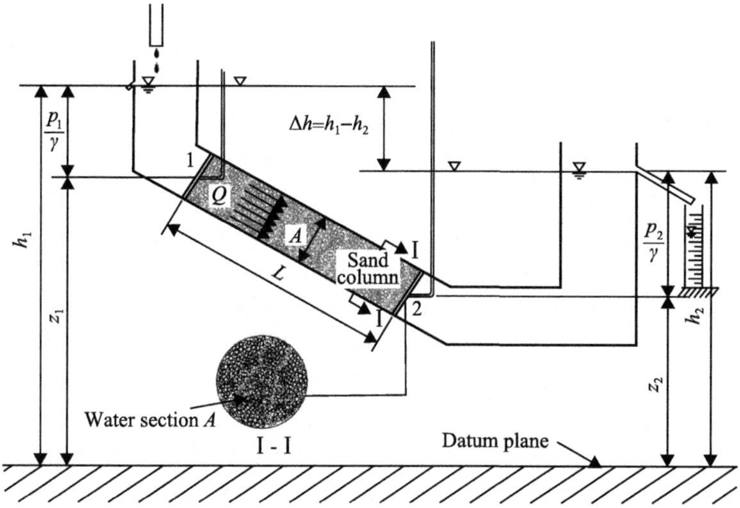

Fig. 3.2 Model of Darcy's experiment

Hydraulic conductivity, K, is an important physical parameter that indicates how easily fluid can move through the pore space of the material. In Darcy's original experiments, he used sands with different grain sizes. However, hydraulic conductivity can also be determined for solid rocks, since most rocks have some pore space.

The discharge rate Q is a volume flux. The flow questions can be rearranged by dividing the flow by the area to determine the Darcy flow

$$q = \frac{Q}{A} = -K\frac{\mathrm{d}h}{\mathrm{d}x} \tag{3.2.3}$$

Or in terms of intrinsic permeability and pressure gradient

$$q = -\frac{k}{\eta}\frac{\mathrm{d}h}{\mathrm{d}x} \tag{3.2.4}$$

where η is the fluid viscosity and k is the intrinsic permeability, and the term $\mathrm{d}h\,/\,\mathrm{d}x$ is the pressure gradient.

It is important to note that the Darcy flux is not equal to the fluid velocity, even though it is expressed in velocity units. The fluid velocity is determined by the average linear velocity (v_a), which is the average of the velocity of all possible fluid paths through the porous medium, but determining this value is beyond the scope of this course. Instead, we find that the fluid velocity is

$$v = \frac{q}{n} \tag{3.2.5}$$

where n is the porosity and v is the fluid velocity. The porosity, n is calculated as ratio of the volume of void space (V_v) to the total volume (V_t) of a material $n = V_v/V_t$, and is usually expressed as a fraction between 0 and 1 or as a percentage. Note that if $n \leq 1$ then the fluid velocity is greater than the Darcy flux.

Another term that appears in the context of flow through rock and unconsolidated material is permeability. While porosity is the actual percentage of pores (voids) in the rock, these voids can have different shapes and connections, which affects how easily a fluid can move through the pore space. Permeability is a measure of the ease with which liquids and gasses can pass through a rock. The more solidified the material, the lower its permeability. Therefore, "loose" materials such as gravel have high permeability. Some rocks may also have anisotropic permeability, meaning liquids can flow easily in one direction but not another. An example of a rock with anisotropic permeability is a shale rock. Liquids can flow easily within one layer of the shale, but not between layers.

From Fig. 3.3, we can see that $v = L/t$ in the absence of granular material. The addition of the pebbles reduces the pore space and thus the porosity, and each fluid particle must travel a longer path to cross the box. The new, longer path created by the addition of the smaller pebbles increases the fluid velocity due to the larger Lcomponent to achieve the same flow rate.

To return to Darcy's experiments: What physical process drives the flow of liquid through the flume?

The term $\mathrm{d}h/\mathrm{d}x$ in Eq. (3.2.3) is called hydraulic head, which is the fluid pressure due to the weight of a fluid relative to a given reference location: a pressure gradient. However, $\mathrm{d}h/\mathrm{d}x$ does not have the unit of a pressure per unit length, so why is it called hydraulic head?

Let' us consider how high the pressure in the gutter is in Darcy's experiments. To conduct the experiments, Darcy had a water tank at the top of the flume to ensure a constant water level at the top of the flume equal to the height of the flume,h_1. Similarly, at the bottom was a container with a constant water level equal to h_2. The fluid pressure in each of the tanks due to the weight of the water is thus given by

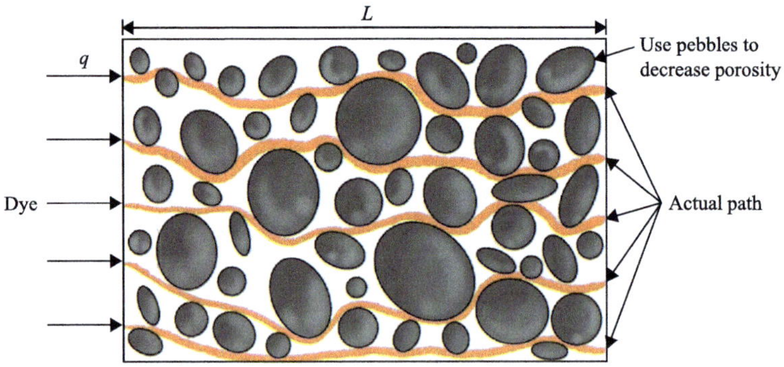

Fig. 3.3 Illustration of actual flow path in the pore space

$p = \rho g h$. The pressure gradient between the upper and the lower end of the cloud is then

$$p = \rho g \frac{\mathrm{d}h}{\mathrm{d}x} \tag{3.2.6}$$

or

$$\frac{\mathrm{d}h}{\mathrm{d}x} = \frac{1}{\rho g} \frac{\mathrm{d}p}{\mathrm{d}x} \tag{3.2.7}$$

By inserting into Darcy's law we can therefore deduce

$$q = -K \frac{\mathrm{d}h}{\mathrm{d}x} = \frac{K}{\rho g} \frac{\mathrm{d}p}{\mathrm{d}x} \tag{3.2.8}$$

Recall that hydraulic conductivity can be written as a function of intrinsic permeability as follows (γ is the weight of the water)

$$K = k \frac{\gamma}{\eta} \tag{3.2.9}$$

which is equivalent to $k(\rho g/\eta)$ or rearranging,

$$\frac{K}{\rho g} = \frac{k}{\eta} \tag{3.2.10}$$

Therefore, we can replace two equivalent expressions for Darcy's law by one expression for the height gradient,

$$q = -K \frac{\mathrm{d}h}{\mathrm{d}x} \tag{3.2.11}$$

and the second in terms of the gradient in pressure,

$$q = -\frac{k}{\eta} \frac{\mathrm{d}p}{\mathrm{d}x} \tag{3.2.12}$$

Thus, $\mathrm{d}h/\mathrm{d}x$ is called the hydraulic head because the physical origin of the term in Darcy's Law is the pressure gradient in the fluid, which drives the flow.

3.2.2 *Cubic Law*

The simplest model for flow through a rock fracture is the parallel plate model. This is the only fracture model for which an accurate calculation of hydraulic conductivity

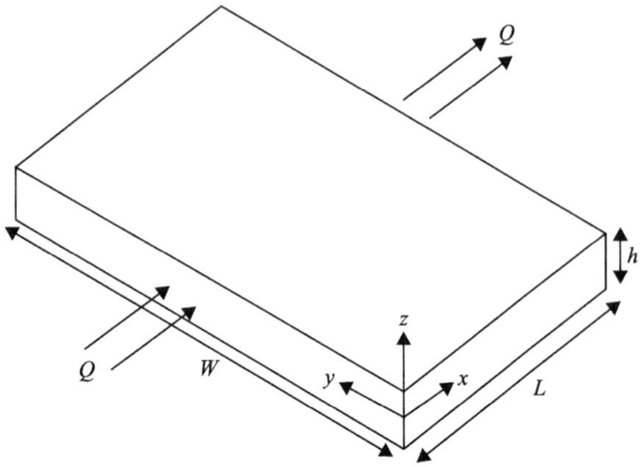

Fig. 3.4 Parallel plate flow model for fractures

is possible; this calculation leads to the well-known cubic law. The derivation of the cubic law begins with the assumption that the fracture walls can be represented by two smooth, parallel plates, which are separated by an opening h (Fig. 3.4). Flow occurs in the space between these parallel plates, from inlet to outlet, as marked by the arrows, with a commonly used boundary condition: Constant static pressures at the inlet and outlet. The flow space otherwise remains bounded by impermeable and rigid fracture walls (slip-free boundary conditions). The width of the fracture is expressed as W and the distance between the inlet and outlet (length of the fracture) is L. This system produces a uniform pressure gradient that is entirely in the plane of the fracture, resulting in unidirectional flow through the system. In this case, the flow is in the x-direction, so only the x-component of the velocity is recorded, u, exists (i.e., $v = 0$; $w = 0$).

The problem of laminar flow of a viscous incompressible fluid in a fracture has been studied by many researchers, starting with Bear (1972). Few of the researchers have derived the basic equations describing the flow through a fracture. If the flow is laminar and the analogy of parallel flat plates is assumed to represent the fracture surfaces, these researchers have shown that the hydraulic conductivity of a fracture with an opening 2b is given by the following equation (see the "Notation" section for definitions of all terms)

$$K_{\mathrm{f}} = (2b)^2 \rho g \,/\, (12\mu) \tag{3.2.13}$$

where μ is dynamic viscosity.

If the flow is uniform and isothermal, the flux per head can be derived from Darcy's law and written in a simplified form as

$$Q \,/\, \Delta h = C(2b)^3 \tag{3.2.14}$$

where C is a constant, which in the case of radial flow is given by

$$C = \frac{2\pi}{\ln\left(r_e/r_w\right)} \frac{\rho g}{12\mu} \tag{3.2.15}$$

and in the case of straight flow by

$$C = \left(\frac{W}{L}\right)\left(\frac{\rho g}{12\mu}\right) \tag{3.2.16}$$

where r_e is radins of influence, r_w is the radins of well.

Equation (3.2.16) is the basis for what is often called the "cubic law" for flow in a fracture. It must be remembered that this equation was derived for an "open" fracture, i.e., the plane surfaces remain parallel and therefore do not touch at any point.

The law for flow in fractures can be generalized using the Reynolds number *Re*. These researchers have shown that the introduction of the factor *D*, then

$$Re = Dv\rho/\mu \tag{3.2.17}$$

$$\psi = \frac{D}{v^2/29}(\nabla h) \tag{3.2.18}$$

The cubic law for laminar flow in an open fracture then reduces to the simple relationship

$$\psi = 96\,/\,Re \tag{3.2.19}$$

The first comprehensive work on flow through open fractures used parallel glass plates and showed the validity of the cubic law as long as the flow is laminar. He also studied the effects of changing from smooth to rough fracture walls and eventually to models with different fracture shapes. He introduced the concept of defining roughness $\in$ in terms of the absolute height of the asperities, and developed the empirical equation

$$\psi = \frac{96}{Re}\left[1 + 6.0\left(\frac{\in}{2b}\right)^{1.5}\right] \tag{3.2.20}$$

which is valid for $\frac{\in}{2b} > 0.065$. Lomize (1951) also considered the effect of flow through fractures with planar but nonparallel (converging or diverging) sides.

We simplify Eq. (3.2.20) by reformulating the equation in the form

$$\psi = \frac{96}{Re} f \tag{3.2.21}$$

where f is a factor that accounts for deviations from the ideal conditions that were assumed in deriving in the case of roughness, $f \geq 1$. The cubic law then becomes

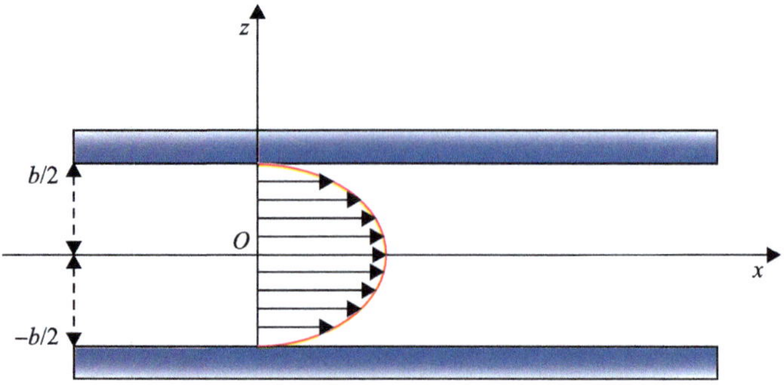

Fig. 3.5 Parallel plate flow model in 2D and the parabolic velocity profile

$$\frac{Q}{\Delta h} = \frac{C}{f}(2b)^3 \tag{3.2.22}$$

Consider a steady, two-dimensional viscous flow between two parallel plates arranged as shown in Fig. 3.5. In fractured media, the permeability of the rock matrix is generally very low. Thus, it is assumed that the rock matrix is an impermeable layer. Therefore, the main channels of water flow are in fractures. When the water flows in laminar flow in the fracture, the fracture wall is smooth and the fracture width(e.g., the fracture opening) is the same everywhere. This ideal fracture model is laminar flow considering the viscous force of the fluid, usually referred to as Poiseuille flow. The velocity distribution in the z-direction (i.e., the fluid velocity as a function of z) can be derived from,

$$u\,(z) = -\frac{b^2}{8\mu}\frac{\mathrm{d}P}{\mathrm{d}x}\left(1 - \frac{4z^2}{b^2}\right) \tag{3.2.23}$$

where b is the fracture opening. It can be seen that the flow velocity in the direction of the fracture opening is parabolic, which is different from the velocity distribution of an ideal fluid (independent of the viscosity of the fluid). The flow velocity of an ideal single fracture can be determined by integrating the flow velocity in the direction of the fracture opening

$$q = \int_{-b/2}^{b/2} u\,(z)\,\mathrm{d}z = -\frac{b^3}{12\mu}\frac{\mathrm{d}P}{\mathrm{d}x} \tag{3.2.24}$$

Note that this flow is directly proportional to the effective pressure gradient and inversely proportional to the fluid viscosity, increasing with the cubic number of the distance between the plates. The total volume flow through the crack for a width W is determined by integrating the velocity across the crack from $z = b/2$ to $z = -b/2$.

Let x be a longitudinal coordinate measuring the distance along the plates, and let z be a transverse coordinate such that the plates are at $z = b/2$ and $z = -b/2$. Assume that there is a uniform effective pressure gradient in the x-direction such that

$$\frac{\mathrm{d}P}{\mathrm{d}x} = -G \tag{3.2.25}$$

where G is a constant. Here, the quantity G could represent a gradient of actual fluid pressure, a gradient of potential gravitational energy (due to the inclination of the plates to the horizontal), or a combination of both, which makes no difference to the final result. Let us assume that the velocity profile of the fluid between the plates has the following form.

3.3 Non-linear Flow Law

Darcy's law, which expresses a linear relationship between velocity and slope, played an important role in the analysis of problems related to porous media for many years. This law was postulated on the basis of experiments carried out under special conditions. These conditions were: ① fully saturated porous medium of medium sand; ② constant temperature; ③ inert porous medium; ④ stable sand bed; ⑤ flow under medium gradient; and ⑥stable laminar flow. Some scientists have shown that Darcy's law can be derived from the Navier–Stokes equations of fluid flow only under these very restrictive conditions.

At low and high gradients, the validity of Darcy's law becomes questionable. The experimentally observed non-Darcy behavior of seepage flows is shown in Fig. 3.6. The three types of flow regimes are prelinear, linear, and postlinear. Molecular forces and inertial forces are assumed to be the main causes of deviations from Darcy's law in the prelinear and postlinear regimes, respectively. Therefore, at low and high gradients, a more accurate relationship between velocity and gradient must be exploited to obtain realistic solutions. Several authors have proposed nonlinear relationships between velocity and gradient at low and high gradients, but the equations of Izbash and Forchheimer are most commonly used.

3.3.1 Izbash Equation

Izbash gave an equation of the form:

$$v = Mi^n \tag{3.3.1}$$

where M and n are the coefficients determined by the fluid flow and medium properties. When $n = 1$, the equation is identical to the Darcy' equation. During the

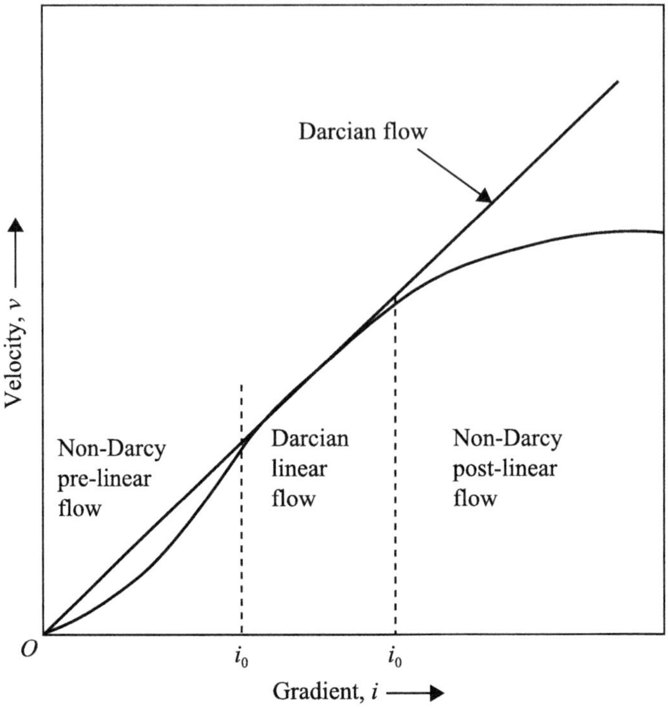

Fig. 3.6 Schematical representation of non-Darcian velocity-gradient response

experiments, it was found that the coefficients M and n are not constant for a material size under all flow conditions, but have different values for different flow regimes. Equation (3.3.1) is entirely empirical and has not been theoretically justified by any researcher; therefore, it is valid only within the limits of the data. In other words, predictive extrapolation beyond the actual test values is generally not advisable.

3.3.2 Forchheimer Equation

Although the Darcy's equation is an empirical equation based on experimental studies, later researchers provided the theoretical background of the equation. In general, Darcy's Law is believed to apply to flows in the laminar region (low Reynolds number). In some flow situations, such as near a well, flowing through water filters, or flowing through rockfill dams, where the Reynolds number is high, the linear relationship between V and i is no longer valid. This could be due to the fact that in a flow with lower Reynolds number, the drag is mainly due to the viscous force. As the Reynolds number increases, the effect of viscous drag decreases and the inertial force begins to dominate, and when the flow becomes completely turbulent, the inertial force becomes the main drag force. Taking into account the inertia effect, a number of nonlinear relations have been proposed over the years by various researchers to simulate the flow in a non-Darcyian domain.

These types of equations are mainly in the form of

$$i = aV + bV^2 \tag{3.3.2}$$

where i is the hydraulic gradient, V is the seepage velocity, a and b are the Darcy and non-Darcy coefficients, respectively.

$$i = \alpha \frac{\mu}{g} \frac{(1-n)^2}{n^3 d^2} V + \beta \frac{1-n}{n^3} \frac{1}{gd} V^2 \tag{3.3.3}$$

where n is porosity and g is acceleration due to gravity, α and β are coefficients related to the pore structure.

Equation (3.3.2), known as the "Forchheimer equation," was originally proposed by a German researcher named P. Forchheimer. Although the equation is based on the analogy of pipe flow, the Forchheimer equation can also be derived from the Navier–Stokes equation. It can be seen from Eq. (3.3.2) that aV represents the energy loss in the laminar region, while bV^2 represents the energy loss in the fully developed turbulent region. Later, the equation was modified several times by many researchers. Forchheimer himself modified the equation by adding a third term to account for the transition conditions.

$$i = aV + bV^2 + cV^3 \tag{3.3.4}$$

Later Eq. (3.3.4) was further modified as

$$i = aV + bV^{1.5} + cV^2 \tag{3.3.5}$$

In Eqs. (3.3.4) and (3.3.5), c is an empirical constant. Equation (3.3.4) was later generalized by Polubarinova-Kochina by adding a time-dependent term as

$$i = aV + bV^2 + c\frac{\mathrm{d}V}{\mathrm{d}t} \tag{3.3.6}$$

More recently, the Forchheimer equation has been modified for lower velocities. It was found that the curve between i/V (hydraulic gradient/velocity) and V (velocity) is not a continuously increasing straight line, but has a negative slope up to a certain velocity value. Although Eqs. (3.3.4), (3.3.5), and (3.3.6) are modified versions of Eq. (3.3.2) and represent all three regimes laminar, transition, and turbulent McCorquodale's Eqs. (3.3.2) and (3.3.4) give the same value of i for Reynolds number 600–4000.

3.4 Multiphase Flow

3.4.1 Basic Concept of Multiphase Flow

1. Saturation

When the void space of a fractured medium is occupied by two or more fluids, the fraction of void occupied by each fluid is a very important parameter. The saturation of a phase fluid at any point in the medium is defined as the percentage of the void volume occupied by the phase fluid in the characteristic voxel around point.

$$S_\alpha = \frac{\Delta V_\alpha}{\Delta V_n} \tag{3.4.1}$$

$$\sum_\alpha S_\alpha = 1 \tag{3.4.2}$$

where ΔV_α is volume of fluid α, ΔV_n is the volume of space.

The cavity is usually filled with two liquids, e.g. a wetting liquid (wetting phase) and a non-wetting liquid (non-wetting phase).

$$S_{\mathrm{w}} + S_{\mathrm{nw}} = 1 \tag{3.4.3}$$

For example, the value of oil saturation in the formation varies widely, generally between 55 and 95%. During drilling, oil saturation in the formation can be determined using data from the rock survey, fluorescence analysis, and returned mud measurement data, or using field geophysical data.

In the laboratory, the method to directly measure the liquid saturation of each phase without destroying the porous medium saturated with different liquids is currently an important research topic. In the early days, the resistance method and the X-ray absorption method were mainly used. In recent years, the ultrasonic method, the radiation detection method and the nuclear magnetic resonance method have been developed. Which method is used depends not only on the instrument conditions, but also on the shape and geometric size of the sample. For thinner samples, the ultrasonic method can be used; if the shape and size are suitable, a nuclear magnetic resonance device can be used for measurement; for larger samples, high-dose radiation should be used for measurement.

2. Contact angle

The contact angle (also called the wetting angle) occurs when a drop of liquid is applied to a material surface and the drop forms a dome shape on the surface. The angle between the surface and the line tangent to the edge of the water droplet is called the contact angle. The contact angle is a measure of the ability of a liquid to wet the surface of a solid. The shape a drop takes on a surface depends on the surface tension of the liquid and the nature of the surface. At the boundary between the droplet and the gaseous environment, the surface tension causes a curved contour.

In Fig. 3.7, θ is the contact angle. As the water drop spreads over a surface and the dome becomes flatter, the contact angle becomes smaller. When the water droplet beads on the surface (like on a water repellent garment or a waxed car), the dome becomes higher and the angle becomes larger. The figure above shows the contact relationship between two immiscible liquids and a solid surface. The state of equilibrium in this situation requires

$$\cos\theta = \frac{\sigma_{\mathrm{SG}} - \sigma_{\mathrm{SL}}}{\sigma_{\mathrm{GL}}} \tag{3.4.4}$$

The angle between the gas–liquid interface and the solid surface is defined as the contact angle θ, which is the angle between the tangent direction of the interface and the solid surface. It is defined as follows: Starting from the tangent of the liquid interface through the tangent of the denser liquid inside to the solid surface. We call this fluid the wetting of the solid. For this solid, this fluid is called the wetting fluid, and the two have an affinity. If the fluid on the solid surface is slightly elliptical, the fluid is called a non-wetting fluid for the solid, and there is hydrophobicity between the two (as shown in Fig. 3.8). Figure 3.9 shows the liquid at various contact angles.

Figure 3.10 shows the fluid distribution of water-wetted rock and oil-wetted rock in different cases. The dashed part in the figure represents the solid rock particles. Figure 3.10a–c show water-wetted sand. Figure 3.10a: water saturation is very low. The water only forms a ring around the contact point of the particles, i.e., a water ring, so it cannot form a continuous water phase. The water ring is in a separated

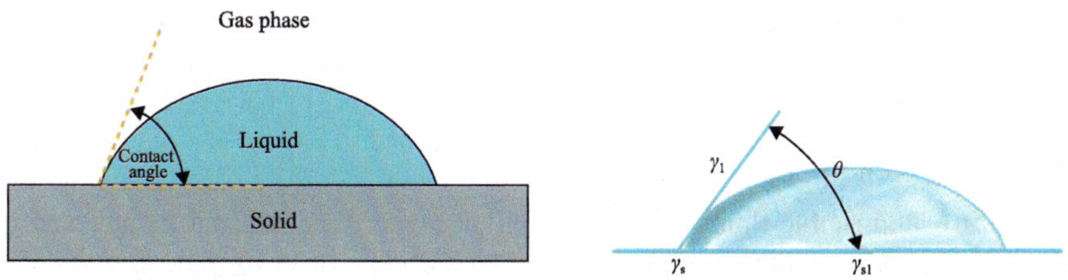

Fig. 3.7 Contact angle

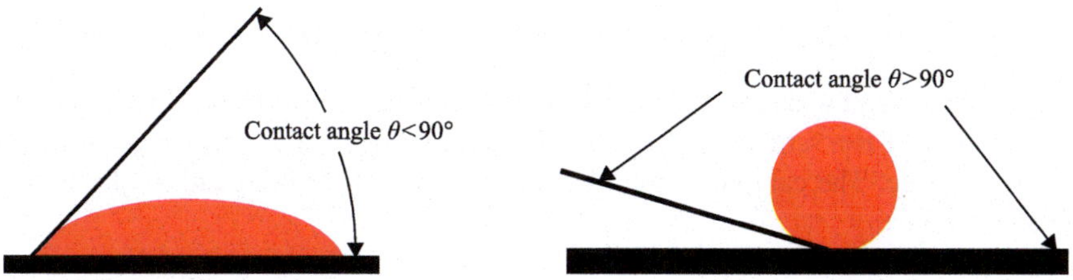

Fig. 3.8 Wetting and non-wetting liquid phase

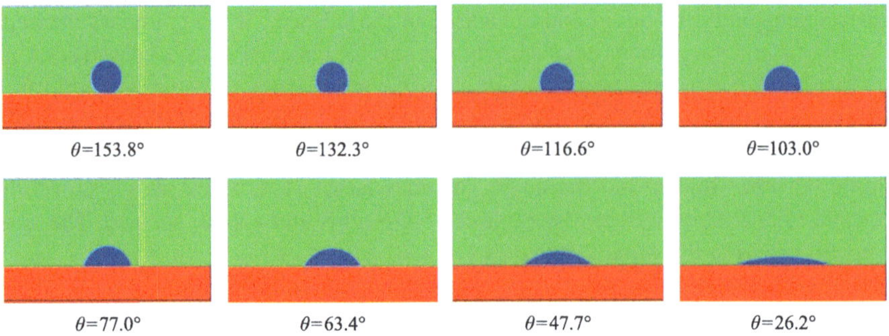

Fig. 3.9 Liquid in different contact angles

state. On the surface of the solid particles there is only a very thin molecular scale water film. In the aqueous phase, the pressure cannot be transferred from one water ring to another, and most of the pore space is occupied by the oil phase. As the saturation of the water increases as the wetting phase, the water ring expands until a continuous wet phase is formed.

The critical saturation at the beginning of the continuous wetting phase is called the equilibrium saturation of the wetting phase. Exceeding this critical saturation is called the equilibrium saturation of the wet phase. At this point, the wet phase can flow, as shown in Fig. 3.10b.

When the saturation of the wet phase increases again, the oil of the nonwet phase breaks down into individual oil droplets instead of forming a continuous phase. These oil droplets are trapped in large pores, as shown in Fig. 3.10c. The oil droplets can flow only if the pressure difference in the pores is large enough.

Figure 3.10d–f show oil-wetted sand. The various conditions are completely similar to those described above, except that the aqueous phase becomes a non-wetting phase

3. Capillary pressure

When two immiscible fluids flow in a capillary, one phase is usually dispersed in the other phase, and many meniscus-like two-phase interfaces form in the region of two-phase flow. The pressure on both sides of the interface is discontinuous, and the difference in this pressure is called the capillary force, which is expressed by p_c

$$p_c = p_{nw} - p_w \tag{3.4.5}$$

where the nw and w denote the non-wetting phase and wetting phase, respectively.

Figure 3.11 is a schematic diagram of capillary force and contact angle for two-phase seepage. The capillary force depends on the radius of curvature of the two-phase interface.

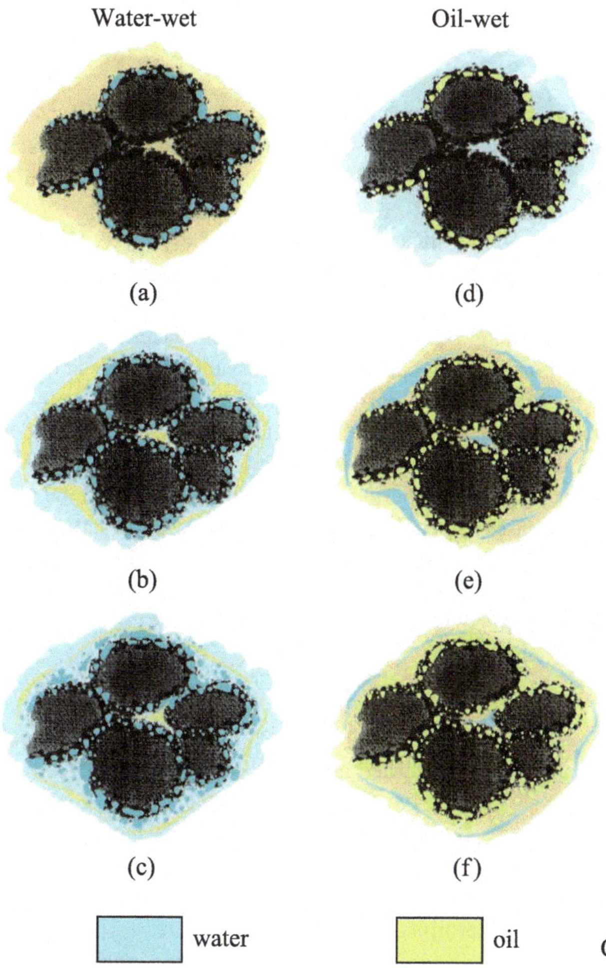

Color picture

Fig. 3.10 Fluid distribution of water wetted rock and oil wetted rock

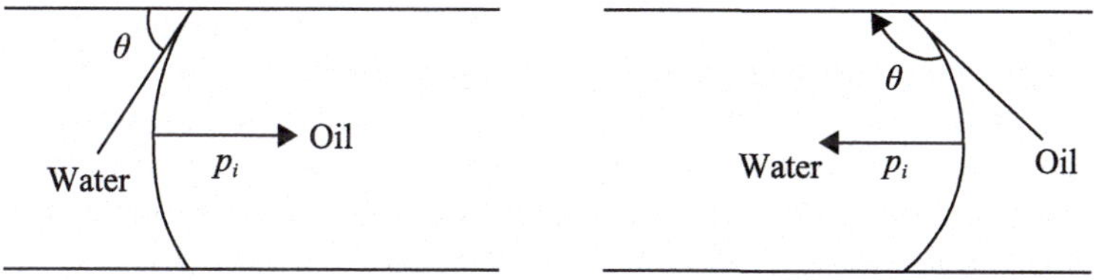

Fig. 3.11 Capillary pressure in two-phase flow

$$p_c = \sigma_{12}\left(\frac{1}{r_1} + \frac{1}{r_2}\right) = \frac{2}{r^*}\sigma_2 \tag{3.4.6}$$

where r_1 and r_2 are the radius of curvature of the two-phase interface, σ_{12} is the surface tension coefficient. Here are the two principal radii of curvature at a given point on the interface, the average radius of curvature (r^*) and the interfacial tension (σ).

The above equation is called Laplaces formula of capillary force. Here are given the two principal radii of curvature at a given point of the interface, the mean radius of curvature, and the interfacial tension. The relationship [Eq. (3.4.6)] can indeed be easily determined by balancing the force on a unit area adjacent to a point on the interface. If the sum is positive, it means that the pressure is greater in the medium with convex surface, i.e., it points to the concave side.

For the actual rock mass, the positions in Eqs. (3.4.5) and (3.4.6) are the average values in the cavity adjacent to the point under study. The capillary force is related to the geometry of the cavity, the properties of solids and liquids, and the saturation of the wet and non-wet phases. Since the shape of the cavity in the actual medium is very complex, it is difficult to describe by an analytical method. For a capillary with radius r, the semi-empirical formula between capillary force, interfacial tension and contact angle can be determined by

$$p_c = \frac{2\sigma}{r}\cos\theta \tag{3.4.7}$$

In practical application, it is concerned about the relationship between capillary force and saturation.

$$p_c = p_c(S_w) \tag{3.4.8}$$

This relationship can be determined experimentally.

The relationship curve between capillary force and saturation is called the capillary force curve, which is determined by Eq. (3.4.7). The experimental results show that the meniscus interface has a contact angle θ when the wet phase fluid and the non-wetting phase fluid are at rest in the capillary. When the wetting liquid displaces the non-wetting liquid, the shape of the meniscus tends to flatten and the contact angle is $\theta_1 > \theta$; when the non-wetting liquid displaces the wetting liquid, the curvature of the meniscus tends to increase and the contact angle is $\theta_2 < \theta$. That is, the contact angle is different when the wetting phase advances or retreats. This phenomenon is observed at the oil–water interface and at the air–water interface. When air displaces water, the contact angle is small; in contrast, the contact angle is large.

Because of this phenomenon, the capillary force curve resulting from the exchange of non-wetting fluid for wetting fluid does not match that resulting from the exchange of wetting fluid for non-wetting fluid. As shown in Fig. 3.12, a test liquid is first fully saturated with wetting liquid and then pressure is applied to the wetting liquid non-wetting fluid, i.e., starting from point *A* in Fig. 3.12, gradually increase or decrease the

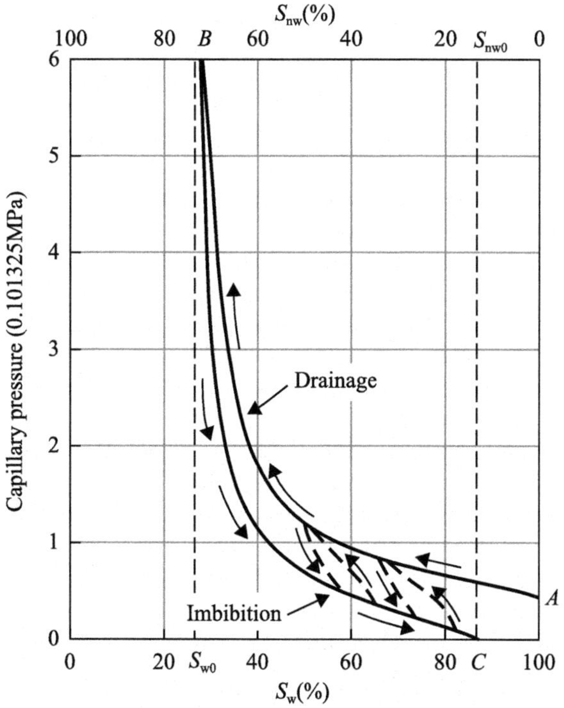

Fig. 3.12 Typical capillary hysteresis in the capillary force curves

pressure of the non-wetting fluid to obtain a capillary force curve, i.e., the outflow curve shown in Fig. 3.12. It can be seen that even with a high capillary force, a certain amount of wetting liquid remains in the sample and is not discharged. The corresponding value is expressed in the so-called irreducible saturation of the wetting liquid.

On the other hand, if S_{w0} is assumed that the wetting liquid is used to displace the non-wetting liquid, which is called imbibition. Unlike precipitation, the wetted liquid can displace the non-wetting liquid as long as it depends on capillary force. When a sample that is fully saturated with a non-wetting liquid is immersed in a wetting liquid, the wetting liquid naturally infiltrates into the sample along the surface of the solid particles and displaces the non-wetting liquid. During the infiltration process, the saturation of the wetting liquid is S_w gradually increases, so we also obtain a $p_c - S_w$ curve. This curve is called the imbibition curve. It can be seen from the figure that this infiltration curve is not along the original curve *AB*, but along a different curve *BC*, and at $p_c = 0$ (point *C* in Fig. 3.12) there is still some non-wetting liquid in the sample. The S_{nw} value of this point is denoted S_{nw0} and is the residual saturation of a non-wetting liquid. If the non-wetting liquid is oil, this is called residual oil saturation. The phenomenon that the outflow curve obtained with the same wetting and non-wetting liquid in the same sample does not coincide with the absorption curve is called capillary hysteresis.

4. Surface tension and wettability

Surface tension is the attractive force exerted on the surface molecules of a liquid by the underlying molecules, which causes the surface molecules to be drawn into the bulk of the liquid and causes the liquid to assume the shape with the least surface area.

Surface tension can be thought of as the rubber of a balloon trying to squeeze the contents into the most compact shape (although this is not quite correct, since in the case of the balloon, the original shape of the balloon can have an effect on the final shape). It is the surface tension that is the "cohesive force" described above. Without gravity and the influence of a solid body coming into contact with the liquid, a drop of water takes on a perfectly spherical shape. This changes as soon as the drop comes into contact with something.

When a liquid w comes into contact with another substance (a liquid, a gas, or a solid that is immiscible with the liquid w), there is an interfacial free energy between them. This interfacial energy is caused by the inward gravitational difference between the molecules in each phase and the molecules at the contact surface, i.e., the molecules in the surface layer store excess free energy due to the imbalance of the molecular force field. When the surface can contract with free energy, the free energy at the interface is expressed as interfacial tension. To separate substances i and k at the contact surface, an external force must do work. The work required to separate a unit area is defined as the interfacial tension σ_{ik}, and its unit is N/m. The interfacial tension between a liquid substance and its own vapor is called the surface tension, which is expressed as σ_{ik}.

Wettability is the ability of a liquid to maintain contact with a solid surface, resulting from intermolecular interactions when the two are brought together. The degree of wetting (wettability) is determined by a balance of forces between adhesive and cohesive forces.

Tthe classical model is a drop of liquid on a smooth solid surface. The contact angle is determined by the balance between the adhesion force (the liquid wants to maintain contact with the solid) and the cohesion force within the liquid (both the internal cohesion force and the surface tension force). An increase in the adhesion force between the liquid and the solid or a decrease in the cohesion force (surface tension) within the liquid will result in greater wettability and a smaller contact angle. It is important to note that a change in adhesion force can be the result of a change in either the solid or the liquid, while a change in cohesion force can only be the result of a change in the cohesion force or the surface tension of the liquid.

5. Relative permeability

In the simultaneous flow of two immiscible liquids through porous media, experimental results show that different liquids form their own tortuous and stable channels. The saturation of the liquid in the wet phase and the liquid in the non-weting phase be S_{w} and S_{nw}, respectively. As the saturation of the non-wetting fluid gradually decreases, the channel of the non-wetting fluid is gradually destroyed, and eventually only some isolated areas retain the residual saturation of the non-wetting fluid. The same is true for the liquid in the wet phase. At S_w gradually decreases, the channel of the wet fluid also becomes damaged and discontinuous when the wet fluid reaches irreducible saturation. When either fluid becomes discontinuous throughout the seepage region, the phase fluid ceases to flow.

To study the simultaneous flow of two immiscible fluids, Darcy's law is extended from the description of the seepage of a single-phase fluid to the seepage of a two-phase fluid. When the two-phase fluid is represented by indices 1 and 2, respectively, it can be written as follows,

$$v_1 = -\frac{k_1}{\mu_1}(\nabla p_1 - \rho_1 g) \tag{3.4.9}$$

$$v_2 = -\frac{k_2}{\mu_2}(\nabla p_2 - \rho_2 g) \tag{3.4.10}$$

where v_1 and v_2 are the seepage velocity of the first and second fluids, and p_1 and p_2 are the fluid pressure of the first and second fluids, ρ_1 and ρ_2 are the density of the first and second fluids; k_1 and k_2 are the intrinsic permeability of the first and second fluids; μ_1 and μ_2 is the dynamic viscosity of the first and second fluids.

Phase permeability is related to the structure of porous media, i.e., to the absolute permeability k of the medium and to the saturation of the phase fluid; it is also related to the properties of another phase fluid that accompanies it. As a rule, one is accustomed to consider its relation to the absolute permeability k

$$k_{r1} = \frac{k_1}{k} \tag{3.4.11}$$

$$k_{r2} = \frac{k_2}{k} \tag{3.4.12}$$

where k_{r1} and k_{r2} are called the relative permeability of Fluid 1 and Fluid 2, respectively. The test shows that for two-phase fluid seepage

$$k_1 + k_2 \neq k \quad \text{or} \quad k_{r1} + k_{r2} \neq 1 \tag{3.4.13}$$

This shows that for the phase permeability of one phase, the other phase cannot be considered as the same group as the medium in the seepage region. In fact, there are some additional forces between two-phase fluids as they cross the medium, and capillary force is one of them. When one phase disperses and moves in the form of droplets or bubbles in the other phase, the radius of the droplets or gas and oil changes from r_2 to r_1 due to the change of the pore diameter in the capillary, and this deformation generates an additional capillary force.

$$\Delta p_c = 2\sigma_{12}\left(\frac{1}{r_2} - \frac{1}{r_1}\right) \tag{3.4.14}$$

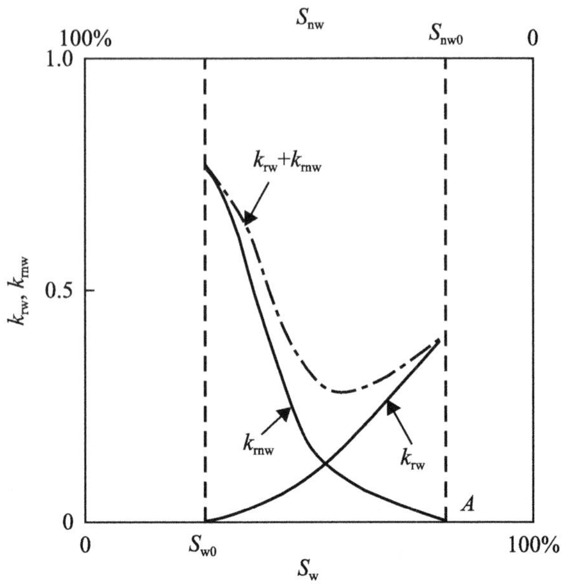

Fig. 3.13 Typical relative permeability curve of oil–water two-phase

This phenomenon is called Jamin phenomenon.

Figure 3.13 is the relationship curve between two typical relative permeabilities and the saturation of the water phase S_w called the relative permeability curve. The figure shows that only when the saturation of the water phase S_w is higher than the irreducible saturation of the water phase S_{w0} and lower than the difference between 1 and the residual saturation of the non-wetting water phase S_{nw0}.

$$S_{w0} < S_w < 1 - S_{nw0} \quad (3.4.15)$$

The two fluids can flow at the same time. Generally speaking, in the state of S_{nw0}, k_{rw} is much smaller than 1; in the state of S_{w0}, k_{rnw} is close to 1. In Fig. 3.13, k_{rw} decreases as S_w decreases rapidly, indicating that the non-wetting fluid first occupies a large void. With the increase of non-wetting phase saturation S_{nw}, the average diameter of voids occupied by wetting phase fluid decreases gradually.

We can roughly divide the relative permeability curve into three regions: In the region of $S_w < S_{w0}$, water is surrounded by a thin film on the surface of solid particles, while oil is in a circuitous continuous distribution, and water has little effect on oil flow. Therefore, the relative permeability of oil is large and decreases slowly.

In the region of $S_{w0} < S_w < 1 - S_{nw0}$, oil and water establish tortuous flow channels respectively. The relative permeability of oil decreases rapidly and that of water increases gradually.

In the region of $S_w > 1 - S_{nw0}$, the oil phase becomes isolated drop distribution and no longer flows, and its relative permeability decreases to zero. The change process of relative permeability of gas–water two-phase or gas-oil two-phase is generally similar to the above.

3.4.2 Immiscible Fluid Flow

Immiscible flow is a common phenomenon that occurs in a variety of industrial processes. In the oil and gas industry, the transport of oil from the reservoir to the processing facilities or to the onshore refinery usually occurs in a multiphase flow because water and oil are normally produced together. The water fraction is usually affected by the presence of water in the reservoir and by the oil production method, which uses water to pump the remaining oil in the reservoir. The presence of water in the transport of oil has a significant impact because the flow can no longer be treated as a single-phase flow. Oil and water have a complex interfacial structure that makes hydrodynamic prediction of fluid flow difficult. Changes in water content can affect the power required to pump the fluid, as the pressure drop in the pipeline changes accordingly. Both water-in-oil and oil-in-water dispersions can drastically affect the pressure drop.

The main methods for studying immiscible two-phase flows are numerical simulation and physical simulation. Only under some simplest conditions approximate analytical solutions can be obtained. For the actual problem of immiscible two-phase flow, under certain boundary conditions, it is necessary to determine the following dependent variables at each point of the rock and at each instant of time:

(1) Pressure p_1, p_2 of two fluids and the capillary force p_c.
(2) Saturation S_1, S_2 of two fluids.
(3) Seepage velocity v_1, v_2 of two fluids.
(4) Density ρ_1, ρ_2 of two fluids and the porosity n.
(5) Viscosity μ_1, μ_2 of two fluids.

This usually involves pressure distribution and saturation distribution. To describe this two-phase seepage, there must be corresponding equations. For a general three-dimensional flow there are six equations.

(1) Transport equation

$$v_1 = -\frac{kk_{r1}}{\mu_1}(\nabla p_1 - \rho_1 g) \tag{3.4.16}$$

$$v_2 = -\frac{kk_{r2}}{\mu_2}(\nabla p_2 - \rho_2 g) \tag{3.4.17}$$

(2) Continuity equation

$$\nabla \cdot (\rho v_1) + \frac{\partial (n\rho_1 S_1)}{\partial t} = 0 \tag{3.4.18}$$

$$\nabla \cdot (\rho v_2) + \frac{\partial (n\rho_2 S_2)}{\partial t} = 0 \tag{3.4.19}$$

(3) Saturation equation

$$S_1 + S_2 = 1 \tag{3.4.20}$$

Equations on fluid density, viscosity and solid media

$$\rho_1 = \rho_1(p), \quad \rho_2 = \rho_2(p) \tag{3.4.21}$$

$$\mu_1 = \mu_1(p), \quad \mu_2 = \mu_2(p) \tag{3.4.22}$$

$$n = n\,(p) \tag{3.4.23}$$

The discussion here is limited to isothermal condition. For non-isothermal seepage, they are also related to temperature T.

(5) The relationship equation between pressure difference and capillary force and the relationship equation between capillary force and saturation:

$$p_2 - p_1 = p_c \tag{3.4.24}$$

$$p_c = p_c(S_w) \tag{3.4.25}$$

It has been assumed that the sum of relative permeability is known.

3.4.3 Immiscible Three-Phase Flow

In the unsaturated zone of fractured media, for example, the corresponding migration problem of organic contaminants is actually an immiscible three-phase seepage problem, i.e., water, nonaqueous phase liquid, and gas. The governing equation for isothermal three-phase seepage is as follows,

$$\nabla \cdot \left(\frac{\rho_o k_{ro}}{\mu_o} \nabla p_o \right) = \frac{1}{k} \frac{\partial(\rho_o n S_o)}{\partial t} \tag{3.4.26}$$

$$\nabla \cdot \left(\frac{\rho_w k_{rw}}{\mu_w} \nabla p_w \right) = \frac{1}{k} \frac{\partial(\rho_w n S_w)}{\partial t} \tag{3.4.27}$$

$$\nabla \cdot \left[\left(\frac{\rho_g k_{sg}}{\mu_g} + \frac{\rho_o k_{ro} R_s}{\mu_o} + \frac{\rho_w k_{rw} R_{sw}}{\mu_w} \right) \nabla p_g \right] = \frac{1}{k} \frac{\partial}{\partial t} \left[n(\rho_o R_s S_o + \rho_w R_{sw} S_w + \rho_g S_g) \right] \tag{3.4.28}$$

where R_s is the ratio of dissolved non-aqueous phase liquid to gas, and is the volume of gas under standard conditions contained in the volume of non-aqueous phase liquid after ground degassing per m^3. R_{sw} is the ratio of the volume of gas escaping from water to the volume of water under standard conditions, that is, the solubility of gas in water. In fact, in the above equation, removing the term containing subscript w is the governing equation for the seepage of non-aqueous liquids and gases in two phases.

In principle, the above equations complement the saturation equation and the capillary force equation, which can be solved under certain boundary conditions and initial conditions. In general, they can only be solved by a numerical method.

References

Bear J (1972) Dynamics of fluids in porous materials. Society of Petroleum Engineers, Richardson

Lomize GM (1951) Flow in fractured rocks. Gesemergoizdat, Moscow

Chapter 4
Basic Process of Mass Transfer in Fractured Media

Mass transfer in fractured media can occur by molecular diffusion and advection. Molecular diffusion is mass transport on a microscopic scale due to a difference in concentration, i.e., from an area of high concentration to an area of low concentration. In accordance with the transport properties of miscible impurities, this chapter mainly presents the basic mass transfer processes in saturated fractured media.

4.1 Diffusion

The process of molecular diffusion is the result of the random movement of dissolved molecules in water. Macroscopically, the process of molecular diffusion is the migration process of solutes in water from a high concentration to a low concentration under the action of a concentration gradient. Whether the groundwater is in motion or not, as long as there is a concentration gradient, the molecular diffusion process will occur. Even if the macro concentration gradient is zero, it does not mean that the random movement of solute molecules in water stops, but a large number of solute molecules move randomly to establish the macro concentration equilibrium.

Let us take the one-dimensional molecular diffusion process as an example and assume that the molecules move randomly only along the x direction. In Fig. 4.1, the concentration is defined as the number of molecules dissolved space size, and B–B' is the concentration interface. Within the time interval Δt, the probability of random left or right movement of each solute molecule is assumed to be equal to $\pm\Delta x$. Within the time interval Δt and within the range of Δx from the concentration interface B–B', the probability of each solute molecule passing through B–B' is 50%. The number of solute molecules passing B–B' from left to right can be expressed as a positive mass flux, i.e., $+0.5C_1\Delta x A$, where A is the area of the concentration interface B–B'. Similarly, the number of solute molecules passing B–B' from right to left can be expressed as a negative mass flux. Thus, under the effect of the concentration gradient,

Z. Dou et al., *Mass Transfer Dynamics of Contaminants in Fractured Media*,
https://doi.org/10.1007/978-981-99-9187-7_4

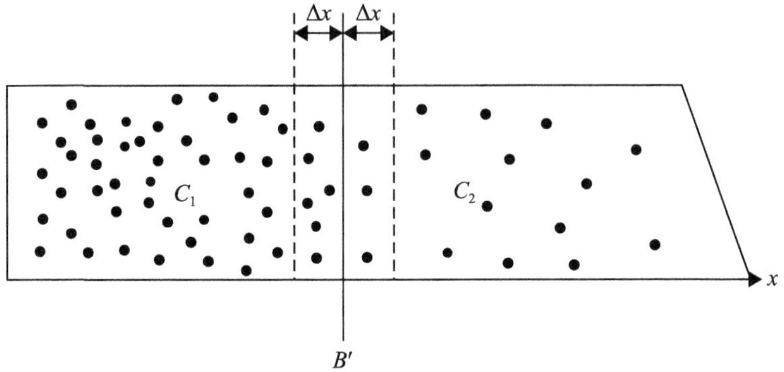

Fig. 4.1 Molecules diffusion process in one dimension

the time interval is Δt and the net physical flux Q of the molecular diffusion process in Fig. 4.1 is expressed as

$$q_x = \frac{+0.5C_1\Delta x A - 0.5C_2\Delta x A}{\Delta t} = \frac{0.5\Delta x A(C_1 - C_2)}{\Delta t} \tag{4.1.1}$$

If the solute concentration $C(x)$ is continuous, then $C_2 \approx C_1 + \Delta x \partial C/\partial x$, the above formula can be transformed into

$$q_x \approx \left(\frac{\Delta x^2}{2\Delta t}\right) A \frac{\partial C}{\partial x} = -DA\frac{\partial C}{\partial x} \tag{4.1.2}$$

In the above formula, D is called the molecular diffusion coefficient, and D is positively correlated with $\frac{(1/2)\Delta x^2}{\Delta t}$, and the dimension is $[\mathrm{L}^2\mathrm{T}^{-1}]$. D indicates how far solute molecules can diffuse in a given solution and in a given time, thus indicating the diffusivity of the solute. The molecular diffusion coefficient depends on the molecular size of the solute, intermolecular polarity, type of solution, and temperature.

Equation (4.1.2) is also known as Fick's law. Fick's law states that the diffusion material flux (i.e., diffusion flux) flowing in a unit time through a unit cross-sectional area perpendicular to the direction of diffusion is directly proportional to the concentration gradient at the cross-section, i.e., the larger the concentration gradient, the larger the diffusion flux. Since the diffusion flux in any direction is orthogonal to the concentration gradient in that direction, Fick's Law in Eq. (4.1.2) can be applied directly to the three-dimensional case,

$$(q_x, q_y, q_z) = \left(-DA_{yz}\frac{\partial C}{\partial x}, -DA_{xz}\frac{\partial C}{\partial y}, -DA_{xy}\frac{\partial C}{\partial z}\right) \tag{4.1.3}$$

Note that the molecular diffusion coefficient in the above formula is assumed to be isotropic. Equation (4.1.3) describes the law of concentration distribution of solute mass at steady state in standing water. Therefore, Eq. (4.1.3) is also known as Fick's

First Law. For a transient system in which the concentration varies with time, Fick's Second Law applies

$$\frac{\partial C}{\partial t} = D\frac{\partial^2 C}{\partial x^2} \tag{4.1.4}$$

Assuming that the diffusion of N solute molecules of total mass m occurs in still water, the random properties of solute molecular motion are considered. For the one-dimensional problem, it is assumed that the mass of each solute molecule is the unit mass and that the molecular motion of the solutes is consistent with Fig. 4.1 (i.e., the probability that the solute molecules move $\pm\Delta x$ at each time step Δt is the same), so that the position of the solute molecules can be predicted by the central limit theory. In finite time, the probability of the position of the solute molecules between $m\Delta x$ and $(m+1)\Delta x$ corresponds to the normal distribution, where the mean of the normal distribution is 0 and the standard deviation is 0. The standard deviation is

$$\sigma = \sqrt{2Dt} \tag{4.1.5}$$

The above formula can be transformed into

$$D = \frac{(1/2)\Delta x^2}{\Delta t} \tag{4.1.6}$$

For a single solute molecule, the probability that the position is between x and $x+\Delta x$ is

$$p(x,t)\Delta x = \frac{1}{\sigma\sqrt{2\pi}}\exp\left(-\frac{x^2}{2\sigma^2}\right)\Delta x = \frac{1}{\sqrt{4\pi Dt}}\exp\left(-\frac{x^2}{2\sigma^2}\right)\Delta x \tag{4.1.7}$$

Thus, if the solute cluster has n solute molecules, at time t, the number of solute molecules located between x and $x + \Delta x$ is,

$$n\,(x,t) = Np(x,t)\Delta x \tag{4.1.8}$$

The solute concentration at position x is expressed as

$$C(x,t) \approx \frac{n(x,t)}{A\Delta x} \tag{4.1.9}$$

The total mass of N solute molecules per unit mass is recorded as M, and Eq. (4.1.9) is transformed into

$$C(x,t) = \frac{M}{A\sqrt{4\pi Dt}}\exp\left(\frac{-x^2}{4Dt}\right) \tag{4.1.10}$$

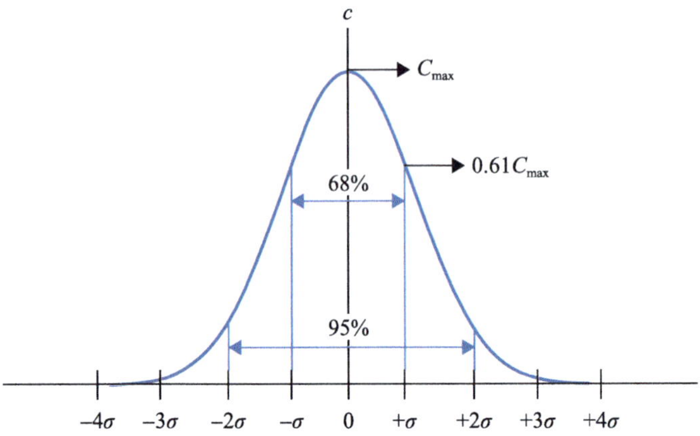

Fig. 4.2 Concentration distribution during the diffusion process of dissolved molecules

As described above, Eq. (4.1.10) shows that the diffusion process of dissolved molecules in water corresponds to the mean 0 and standard deviation $\sigma = \sqrt{2Dt}$ of the normal distribution, as shown in Fig. 4.2. The diffusion process of dissolved molecules is still in water. At any time t, 68% of the solute molecules $\pm\sigma$ are within the range, 95% of the solute molecules are within $\pm$ 2% of the central point of the solute cluster σ within the range, 99.7% of the solute molecules are within $\pm$ 3% of the central point of the solute cluster σ within the range. In general, all solute molecules in the solute cluster are within $\pm\ 4\sigma$ range, i.e., the one-dimensional solute mass length $L = \pm 4\sigma$.

Let us take the one-, two-, and three-dimensional molecular diffusion process for solutes as an example to further illustrate the diffusion process for solutes. According to the equation for conservation of mass, the equation for molecular diffusion of solutes is as follows

$$\frac{\partial C}{\partial t} + u\frac{\partial C}{\partial x} + v\frac{\partial C}{\partial y} + w\frac{\partial C}{\partial z} = D\left(\frac{\partial^2 C}{\partial x^2} + \frac{\partial^2 C}{\partial y^2} + \frac{\partial^2 C}{\partial z^2}\right) \tag{4.1.11}$$

As shown in Fig. 4.3, the solute of mass m is immediately injected into still water at time $t = 0$. Considering the diffusion process of one-dimensional solute molecules, i.e., assuming that the solute concentration is uniform in the y and z directions, Eq. (4.1.11) is transformed into

$$\frac{\partial C}{\partial t} = D\frac{\partial^2 C}{\partial x^2} \tag{4.1.12}$$

The corresponding initial conditions are

$$C(x) = M\delta(x) \tag{4.1.13}$$

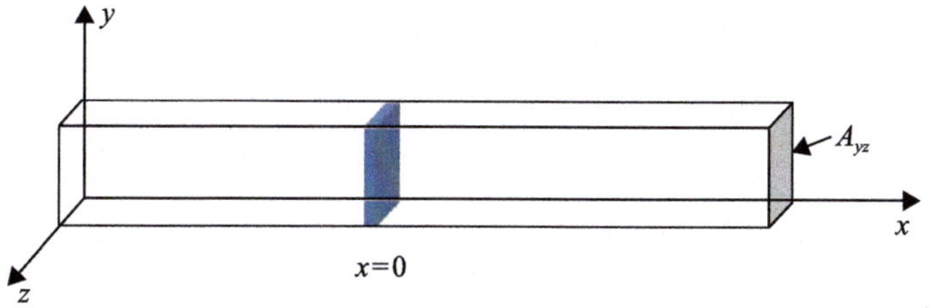

Fig. 4.3 The diffusion process of dissolved molecules

In the above formula $\delta(x)$ is the Dirac function, which is used to represent the instantaneous process of solute injection with mass M. According to Eqs. (4.1.12) and (4.1.13), the analytical solution of the concentration distribution is

$$C(x,t) = \frac{M}{\sqrt{4\pi Dt}} \exp\left(-\frac{x^2}{4Dt}\right) \tag{4.1.14}$$

For the two-dimensional process of molecular diffusion of the solute, it is also assumed that the solute molecule with mass M is immediately injected into still water, and its initial state is

$$C(x, y, t = 0) = M\delta(x)\delta(y) \tag{4.1.15}$$

Considering the anisotropy, the molecular diffusion coefficient is more general, that is, $D_x \neq D_y$. The solute molecular diffusion equation is

$$\frac{\partial C}{\partial t} = D_x \frac{\partial^2 C}{\partial x^2} + D_y \frac{\partial^2 C}{\partial y^2} \tag{4.1.16}$$

The corresponding analytical solutions of Eqs. (4.1.15) and (4.1.16) are

$$C(x, y, t) = \frac{M}{A_{yz} 4\pi t \sqrt{D_x D_y}} \exp\left(-\frac{x^2}{4D_x t} - \frac{y^2}{4D_y t}\right) \tag{4.1.17}$$

For the three-dimensional solute molecular diffusion process, it is also assumed that the solute molecule with mass M is instantaneously injected into still water under the initial condition of

$$C(x, y, z, t = 0) = M\delta(x)\,\delta(y)\,\delta(z) \tag{4.1.18}$$

Considering the anisotropy, the molecular diffusion coefficient is more general, that is, $D_x \neq D_y \neq D_z$. The solute molecular diffusion equation is

$$\frac{\partial C}{\partial t} = D_x \frac{\partial^2 C}{\partial x^2} + D_y \frac{\partial^2 C}{\partial y^2} + D_z \frac{\partial^2 C}{\partial z^2} \tag{4.1.19}$$

The corresponding analytical solutions of Eqs. (4.1.18) and (4.1.19) are

$$C(x, y, t) = \frac{M}{(4\pi t)^{3/2}\sqrt{D_x D_y D_z}} \exp\left(-\frac{x^2}{4D_x t} - \frac{y^2}{4D_y t} - \frac{z^2}{4D_z t}\right) \tag{4.1.20}$$

4.2 Brownian Motion and Fick's Law

4.2.1 Brownian Motion

Brownian motion is a stochastic process that has applications in fields as diverse as economics, biology, and management science. Mathematically, it can be thought of as a continuous-time process in which the unit under consideration moves one "step" in a particular direction at each infinitesimal time interval Δt. This suggests that Brownian motion can be viewed as a "random walk" process, as we will show in a moment.

The physical manifestation of Brownian motion was observed by the Scottish botanist Robert Brown in 1827. His interpretation of this process was based on the movement of small pollen particles suspended in a drop of water. In his experiments, the pollen particles appeared to move completely randomly, which puzzled Brown and his colleagues. Upon further investigation, Brown and others found that this phenomenon occurred not only with pollen particles, but with many different types of microscopic particles suspended in a liquid.

Brown's initial observations of Brownian motion remained without significant mathematical or even physical explanation for nearly 80 years, until Einstein published a paper in 1905 claiming that the motion of these microscopic particles was due to the constant forces exerted on the particles by the fluid surrounding them, thanks to the individual fluid molecules colliding against the particles, causing them to move. Since the pollen grains were completely surrounded by these fluid molecules, they experienced these forces in every conceivable direction, which explained why the particles did not move in any particular pattern or direction.

The actual development of Brownian motion as a stochastic process did not occur until 1923, when Norbert Wiener, a mathematician from MIT, created the modern mathematical framework for what is now known as the random Brownian process. For this reason, Brownian motion is sometimes referred to as the Wiener process. In fact, the study of Brownian motion today is primarily concerned with the stochastic process developed by Wiener rather than the physical process studied by Brown.

We can get an idea of the behavior of Brownian motion by comparing it to simple random motion. In particular, let us assume that a random motion is $X(t)$ increases by an infinitesimal step size Δx with probability p at each infinitesimal time step Δt and decreases by Δx with probability 1–p. This allows us to treat the process $X(t)$ as a continuous function of t where the change in x has no discontinuities. Then we have for each step that

$$\begin{aligned} E[X(t+\Delta t)-X(t)] &= p\Delta x+(1-p)(-\Delta x)=(2p-1)\Delta x \\ E[\{X(t+\Delta t)-X(t)\}^2] &= p(\Delta x)^2+(1-p)(-\Delta x)^2=(\Delta x)^2 \end{aligned} \tag{4.2.1}$$

The number of time units of duration Δt in the time t is $\left[\frac{t}{\Delta t}\right]$, where $\left[\frac{t}{\Delta t}\right]$ is the largest integer that is less than or equal to $\frac{t}{\Delta t}$. Thus, the location of $X(t)$ at time t is given by the sum of $\left[\frac{t}{\Delta t}\right]$ random variables each of which has a value Δx with probability p and a value $-\Delta x$ with probability $1-p$. The expected value of the sum of these random variables is the mean of their sum, which is $\left[\frac{t}{\Delta t}\right](2p-1)\Delta x$. Since these random variables are independent and identically distributed, the variance of their sum is $[t/\Delta t](\Delta x)^2$.

If this process is to approximate the Brownian motion $W(t) \sim N(\mu t, \sigma^2 t)$, we must have that in the limit

$$\begin{aligned} \left[\frac{t}{\Delta t}\right](2p-1)\Delta x &= \mu t \\ \left[\frac{t}{\Delta t}\right](\Delta x)^2 = \frac{(\Delta x)^2}{\Delta t}t &= \sigma^2 t \end{aligned} \tag{4.2.2}$$

If for large t we approximate $\left[\frac{t}{\Delta t}\right]$ by $\frac{t}{\Delta t}$, then from these two equations we obtain

$$\begin{aligned} \Delta x &= \sigma\sqrt{\Delta t} \\ p &= \frac{1}{2}\left(1+\frac{\mu\Delta t}{\Delta x}\right)=\frac{1}{2}\left(1+\frac{\mu}{\sigma}\sqrt{\Delta t}\right) \end{aligned} \tag{4.2.3}$$

Let $T > 0$ be some fixed interval and let $\Delta t = \frac{T}{n}$, where $n = 1, 2, \ldots$. Assume that the random variables Y_k, $k = 1, 2, \ldots, n$, independently take values $\sigma\sqrt{\Delta t}$ with $= \frac{1+\frac{\mu\sqrt{\Delta t}}{\sigma}}{2}$ and $-\sigma\sqrt{\Delta t}$ with probability $1 - \frac{1+\frac{\mu\sqrt{\Delta t}}{\sigma}}{2}$. Define S_n by

$$S_n = \sum_{k=1}^{n} Y_k \tag{4.2.4}$$

Then from the central limit theorem, $S_n \to W(t)$ as $n \to \infty$. Thus, the Brownian motion can be regarded as a random walk defined over an infinitesimally small step size Δx and infinitesimally small time intervals Δt between walks such that both Δx and Δt go to 0 in such a way that $\frac{(\Delta x)^2}{\Delta t}$ remains constant.

More formally, the Brownian motion $\{W(t),\ t \geq 0\}$ is a stochastic process that models continuous random motion. It is considered the continuous-time analog of the random walk and can also be viewed as a continuous-time Gaussian process with independent increments. In particular, Brownian motion has the following properties:

(1) $W(0) = 0$; that is, it starts at 0.
(2) $W(t)$ is continuous in $t \geq 0$; that is, it has continuous sample paths with no jumps.
(3) It has both stationary and independent increments.
(4) For $0 \leq s < t$, the random variable $W = W(t) - W(s)$ has a normal distribution with mean 0 and variance $\sigma_W^2 = \sigma^2(t - s)$. That is, $W - N(0, \sigma^2(t - s))$.

Brownian motion is an important building block for modeling continuous-time stochastic processes. In particular, it has become an important framework for modeling financial markets. The path of a Brownian motion is always continuous, but nowhere smooth; consequently, it is nowhere differentiable. The fact that the path is continuous means that a particle in Brownian motion cannot instantaneously jump from one point to another. Because according to the property (4)

$$W(t) = W(t) - W(0) - N(0, \sigma^2(t - 0)) = N(0, \sigma^2 t) \tag{4.2.5}$$

Thus, $W(t - s) \sim N(0, \sigma^2(t - s))$; that is, $W(t) - W(s)$ has the same distribution as $W(t-s)$. This also means that another way to define a Brownian motion $\{W(t), t \geq 0\}$ is that it is a process that satisfies conditions 1, 2, and 3 along with the condition $W(t) \sim N(0, \sigma^2 t)$.

There are many reasons for studying Brownian motion. As mentioned earlier, it is an important building block for modeling continuous-time stochastic processes, since many classes of stochastic processes contain Brownian motion. It is a Markov process, a Gaussian process, a martingale, a diffusion process, and also a Levy process. Over the years, it has developed into an extensive mathematical object. For example, it is the central object of stochastic calculus.

Brownian motion is sometimes called Wiener process. A sample function of the Wiener process is shown in Fig. 4.4.

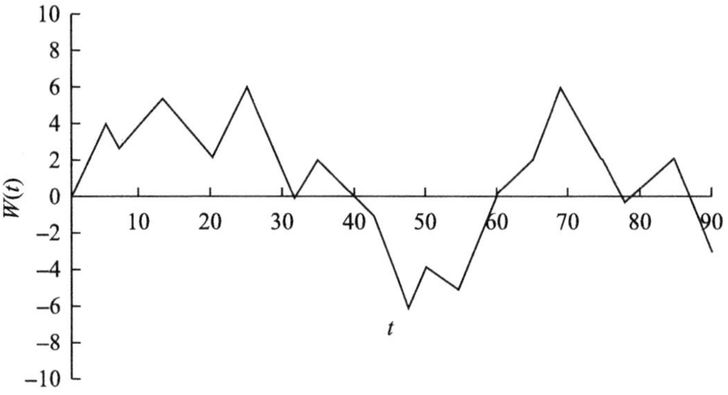

Fig. 4.4 Sample function of a Wiener process

Let $B(t) = W(t)/\sigma$. Then $E[B(t)] = 0$ and $\sigma^2_{B(t)} = 1$. The stochastic process $\{B(t), t \geq 0\}$ is called the standard Brownian motion, which has the property that when sampled at regular intervals it produces a symmetric random walk. Note that $B(t) \sim N(0, t)$. In the remainder of this chapter, we refer to the Wiener process $W(t) - N(0, \sigma^2 t)$ as the classical Brownian motion and use the two terms interchangeably.

4.2.2 Fick's First Law

Equation (4.2.6) is an expression of Fick's First Law of Diffusion, which states that the transfer rate of a diffusing substance through a unit cross-sectional area of the diffusion medium is proportional to the concentration gradient in the direction of transfer.

$$F = \frac{M}{At} = \frac{D(C_1 - C_2)}{1} \tag{4.2.6}$$

The application of this law to simple stationary systems, such as the one shown in Fig. 4.5, is straightforward. However, most practical problems encountered are nonstationary, i.e., the concentration distribution and hence the concentration gradient at all points change with time. Returning to the example of the dye diffusing in a cylinder (see Fig. 4.6), we can represent the process schematically by drawing concentration-distance distributions at arbitrary times (see Fig. 4.7). By "freezing" the process at an arbitrary time t_1 and examining the concentration distribution, we can imagine that it consists of a large number of infinitesimally small sections of thickness dx, which can be treated as the medium F in Fig. 4.5. Thus, at the point x, the concentration gradient is $\frac{\mathrm{d}C}{\mathrm{d}x}$ and at this particular instant of time, t_1, the rate of transfer at x, $F(x)$, is given by

$$F(x) = -D\frac{\mathrm{d}C}{\mathrm{d}x} \tag{4.2.7}$$

In order to acknowledge the fact that the concentration depends not only on x, but also on time, t, Fick's Law should be expressed as the partial differential equation. [Note the change from 'd' to '∂' between Eqs. (4.2.7) and (4.2.8)]

$$F(x) = -D\frac{\partial C}{\partial x} \tag{4.2.8}$$

To solve transient problems, an equation relating concentration to both distance and time is required. Fick's Second Diffusion Law, commonly known as the "diffusion equation," is such a relationship.

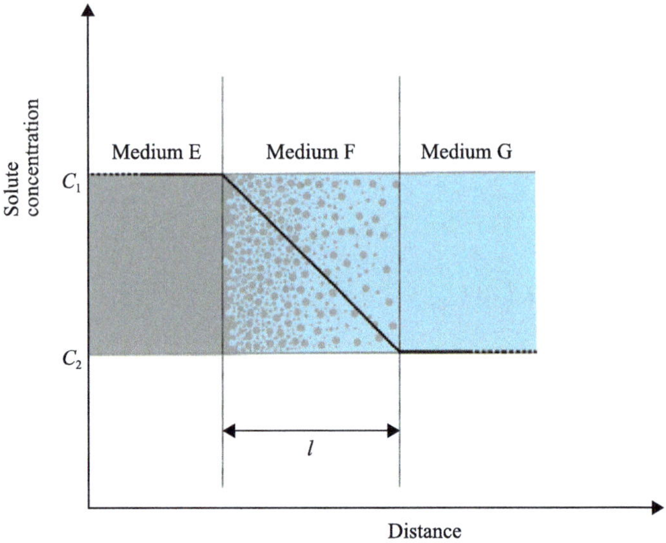

Fig. 4.5 Diffusion in the steady state

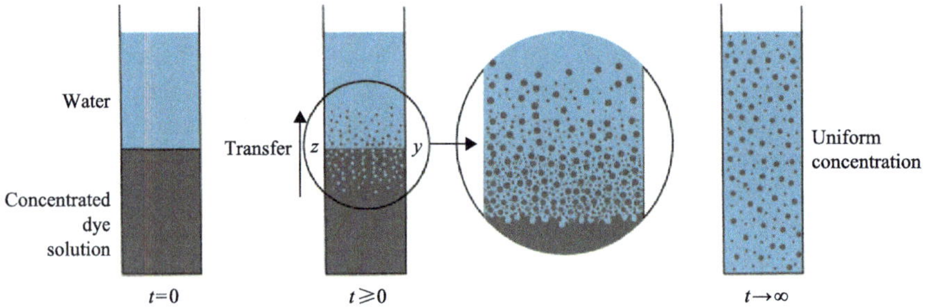

Fig. 4.6 Diffusion of solution of dye into clear water

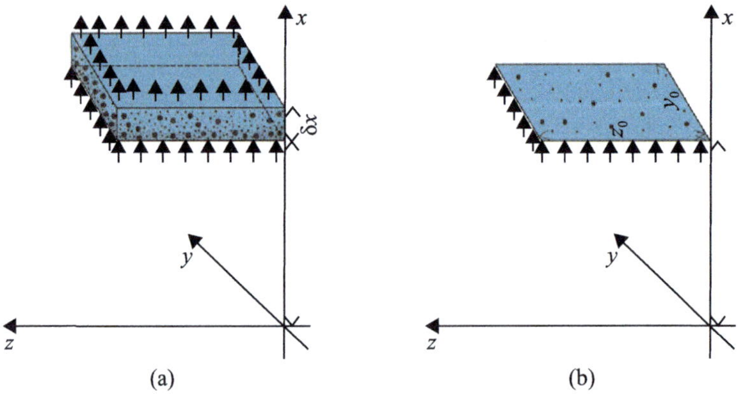

Fig. 4.7 One dimensional diffusion

4.2.3 Fick's Second Law

The flux, F, of diffusing material normal to an area has been discussed earlier and is found to depend upon the area and upon the concentration gradient across that area. If consideration is restricted to flow in a single direction–the x-direction, say–then the flux through an area y_0 by z_0 normal to the flow is proportional to the product, $y_0 z_0$, and to the concentration gradient in the x-direction. The concentration is likely to depend upon both distance x and time t, and so

$$C = C(x, t) \tag{4.2.9}$$

and the concentration gradient is the first partial derivative, $\frac{\partial C}{\partial x}$. Such a flow is shown in Fig. 4.7a. Flow is from a higher to a lower concentration, and so

$$F = -D y_0 z_0 \frac{\partial C}{\partial x} \tag{4.2.10}$$

where D is a constant of proportionality, the diffusion coefficient.

Consider now a second section, parallel to the first, but "downstream" by a small distance, δx, as shown in Fig. 4.7b. (The symbol δ is frequently used in mathematics to indicate "a small part of"). There will be an inflow of diffusing material through the first section and an outflow from the second. Any difference between the two can only be accounted for by a change in the material inside the volume between the two sections. With the flux at the first section being indicated by F_x and at the second by $F_{x+\delta x}$, the mean rate of change of F with respect to x over the distance δx is

$$\frac{F_{x+\delta x} - F_x}{\delta x} \tag{4.2.11}$$

Provided that δx is very small this is very nearly $\frac{\partial F}{\partial x}$. Hence

$$F_{x+\delta x} \approx F_x + \frac{\partial F}{\partial x}\delta x \tag{4.2.12}$$

The net rate of gain of material within the volume is, therefore

$$F_x - F_{x+\delta x} \approx -\frac{\partial F}{\partial x}\delta x \tag{4.2.13}$$

This must equal the alternative expression of volume x rate of change of concentration or

$$y_0 z_0 \delta x \frac{\partial C}{\partial t} \tag{4.2.14}$$

Hence

$$y_0 z_0 \delta x \frac{\partial C}{\partial t} = -\frac{\partial F}{\partial x} \delta x \tag{4.2.15}$$

However, using Fick's First Law (Eq. 4.2.7) to substitute for F gives

$$y_0 z_0 \frac{\partial C}{\partial t} = -\frac{\partial}{\partial x}\left(-D y_0 z_0 \frac{\partial C}{\partial x}\right) \tag{4.2.16}$$

or

$$\frac{\partial C}{\partial t} = \frac{\partial}{\partial x}\left(D \frac{\partial C}{\partial x}\right) \tag{4.2.17}$$

If D is independent of x or C, this becomes Fick's Second Law, or the diffusion equation:

$$\frac{\partial C}{\partial t} = D \frac{\partial^2 C}{\partial x^2} \tag{4.2.18}$$

The same equation appears in problems of heat conduction, where it is known as the heat equation. In such problems, C stands for the temperature and D for the heat transfer coefficient.

4.3 Advection

Dissolved solids are entrained with flowing groundwater. This process is called advective transport or advection. The amount of transported solutes is a function of their concentration in the groundwater and the amount of flowing groundwater. For one-dimensional flow perpendicular to a unit cross-sectional area of the porous medium, the amount of flowing water is equal to the average linear velocity times the effective porosity. Average linear velocity, v_2 is the rate at which the flux of water across the unit cross-sectional area of pore space occurs. It is not the average velocity at which water molecules move along each flow path that is greater than the average linear velocity due to tortuosity. The effective porosity, n_y is the porosity through which flow can occur. Noninterconnected and dead-end pores are not included in the effective porosity, so that $n_e < n$.

$$v_x = \frac{K}{n_e} \frac{\mathrm{d}h}{\mathrm{d}l} \tag{4.3.1}$$

where

$$v_x = \text{average linear velocity } (L/T)$$
$$K = \text{hydraulic conductivity } (L/T) \quad (4.3.2)$$

$$n_e = \text{effective porosity}$$
$$\frac{dh}{dl} = \text{hydraulic gradient } \left(\frac{L}{L}\right) \quad (4.3.3)$$

Note that the abbreviation n is used in the context of flow and transport under saturated conditions. When discussing unsaturated flow, θ is preferred because it refers to the water content of the porous matrix. When the matrix is fully saturated, n equals θ. The one-dimensional mass flux, F_x, due to advection, is equal to the amount of water flowing through times the concentration of dissolved solids and is given by Eq. (4.3.1):

$$F_x = v_x n_e C \quad (4.3.4)$$

The one-dimensional advective transport equation is

$$\frac{\partial C}{\partial t} = -v_x \frac{\partial C}{\partial x} \quad (4.3.5)$$

Solving the equation for advective transport results in a sharp concentration front. On the advancing side of the front, the concentration is equal to that of the infiltrating groundwater, while on the other side of the front it is unchanged from the background value. This is called plug flow, where all the pore fluid is replaced by the intruding dissolved front. The sharp interface created by plug flow is shown in Fig. 4.8. The vertical dashed line at V represents an advancing solute front created by advection alone.

Because of the heterogeneity of geologic materials, advective transport in different layers can cause solute fronts to propagate at different rates in each layer. If a water sample is taken to monitor the spread of a dissolved contaminant from a borehole that penetrates multiple layers, the water sample will be a mixture of the water from each layer. Because advection transports solutes at different rates in each layer, the composite sample may be a mixture of water with the transported solute from one layer and uncontaminated groundwater from another layer where the average linear velocity is lower. The concentration of the contaminant in the composite sample would thus be lower than in the source.

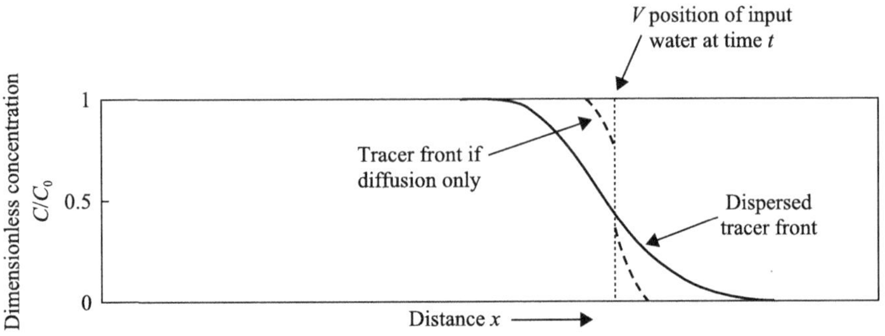

Fig. 4.8 Advective transport and the influence of longitudinal dispersion and diffusion on the transport of a solute in one-dimensional flow

4.4 Difference in Dispersion and Diffusion

Equation (4.4.1) can describe the migration process of solutes in groundwater. In the case of solute migration in nonflowing groundwater, the convection term of Eq. (4.4.1) is 0 and solutes migrate only by molecular diffusion; in the case of solute migration in flowing groundwater, the convection term is not 0 and solute migration occurs by the combined action of molecular diffusion and dispersion. Thus, even if the groundwater is not flowing, the solute migrates by molecular diffusion. The necessary and sufficient condition for the dispersion process of solute migration is that the groundwater is flowable, i.e. that the hydraulic gradient is not 0.

In both porous media and fractured media, the distribution of groundwater flow velocity is very complex, making the flow field description in Eq. (4.4.1) very difficult. Therefore, the flow field velocity in Eq. (4.4.1) is replaced by the mean linear velocity. In an isotropic medium with uniform velocity field, the following equation is obtained for a two-dimensional flow in x-direction

$$D_{\mathrm{L}}\frac{\partial^2 C}{\partial x^2}+D_{\mathrm{T}}\frac{\partial^2 C}{\partial x^2}-\frac{\partial}{\partial x}(\overline{v}_x C)=\frac{\partial C}{\partial t} \tag{4.4.1}$$

where D_{L} is called the longitudinal hydrodynamic dispersion coefficient (L^2/T); D_{T} is called the transverse hydrodynamic dispersion coefficient (L^2/T). $\overline{v}_x$ is the average linear velocity in the x-direction.

Indeed, average linear velocity cannot fully describe the characteristics of groundwater flow. Flowing groundwater moves at velocities greater or less than the average linear velocity, as shown in Fig. 4.9.

When groundwater moves in fractured media, the groundwater often has velocities greater than or less than the average linear velocity due to the non-uniform shape and size of the fractured channels: ① The distribution of crack width has a significant effect on groundwater velocity; ② For the same straight-line distance, the migration paths of groundwater flow particles are not exactly the same, and some fluid particles

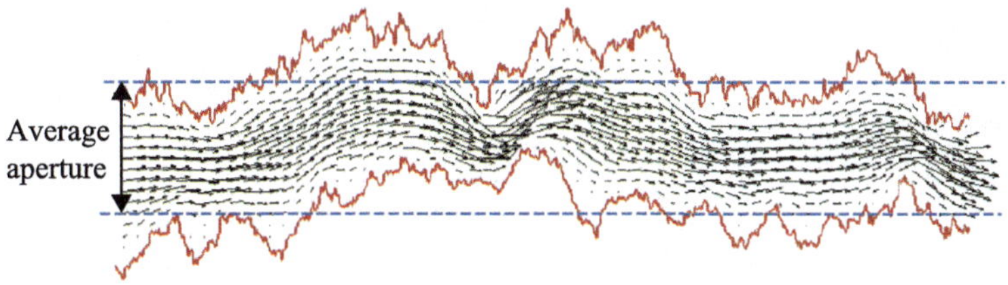

Fig. 4.9 Distribution of groundwater flow field in rough fractured medium

migrate along a longer path than other particles; ③ The velocity at the center of the crack is greater than that at the crack wall.

Figure 4.10 shows the solute transport process based on the distribution of the flow field in Fig. 4.9. Because of the difference in velocity distribution of the subsurface fluid in the crack media, the subsurface fluid containing the solutes does not migrate at the same velocity, and the convective migration of the solutes in each crack is not uniform. The convective migration caused by this difference in velocity results in mechanical mixing of the solutes, which is called mechanical dispersion. When this mechanical mixing occurs along the direction of convection, it is macroscopically referred to as longitudinal mechanical dispersion. At the pore scale, the flow path diverges, and the convection-driven solute front tends to expand along the vertical flow direction, as shown in Fig. 4.11. This mechanical mixing is referred to as transverse mechanical dispersion.

To quantify the phenomenon of mechanical dispersion, assume that mechanical dispersion can be described by Fick's diffusion law (Eq. 4.2.6). After replacing the actual differential fluid velocity in the subsurface with the average linear velocity, the flux of mechanical dispersion is considered to be a function of the average linear velocity. In terms of Fick's diffusion law, we can introduce the mechanical dispersion coefficient. In the longitudinal and transverse directions,

$$\text{Longitudinal mechanical dispersion coefficient} = \alpha_i v_i$$
$$\text{Transverse mechanical dispersion coefficient} = \alpha_j v_i$$

As mentioned earlier, the solute migrates by molecular diffusion even if the groundwater is not flowing. As long as there is a concentration gradient, the molecular diffusion process occurs separately. Therefore, the molecular diffusion process cannot be separated from mechanical dispersion when the groundwater is flowing. In other words: When solutes actually migrate, mechanical dispersion and molecular diffusion occur simultaneously. The interaction of these two processes causes the solutes in the groundwater to migrate. The combination of mechanical dispersion and molecular diffusion is called a hydrodynamic dispersion process. Define the corresponding hydrodynamic dispersion coefficient.

$$D_{\mathrm{L}} = \alpha_{\mathrm{L}} v_i + D_{\mathrm{m}} \tag{4.4.2}$$

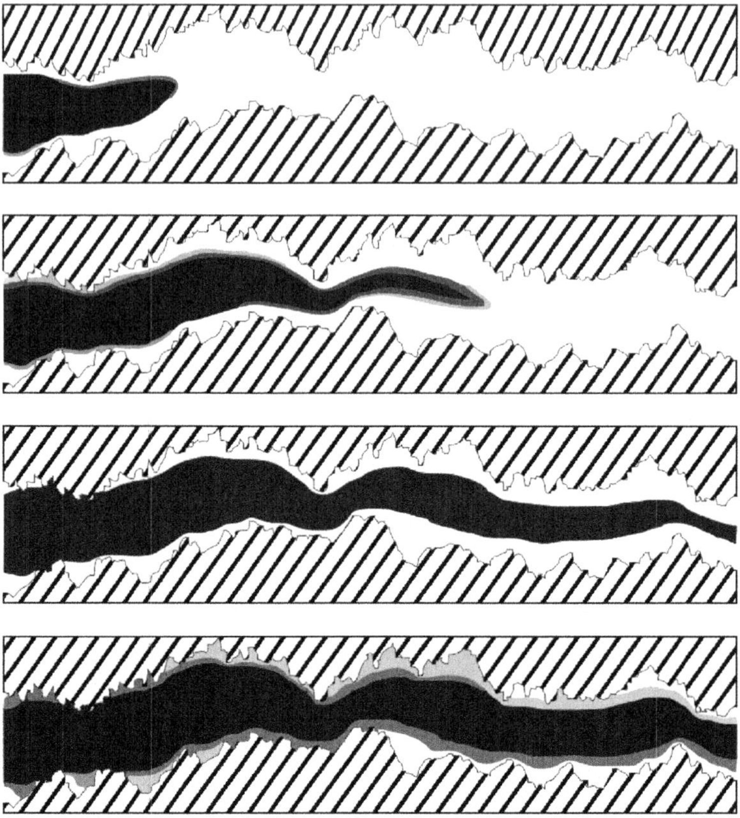

Fig. 4.10 Solute transport process in rough fractured media at different times

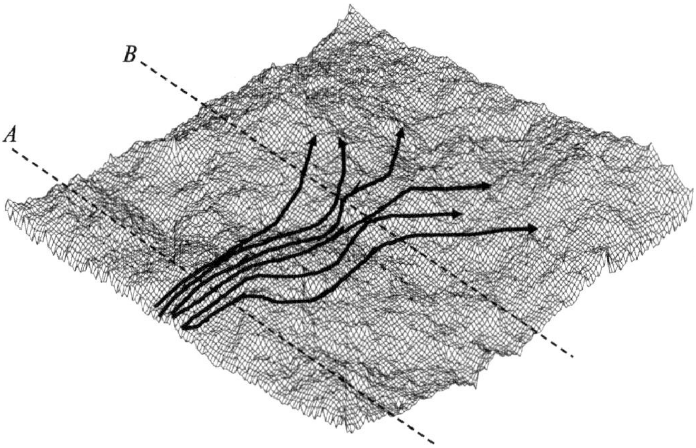

Fig. 4.11 Streamline leading to transverse hydrodynamic dispersion of solute in rough fracture structural planes

$$D_T = \alpha_T v_i + D_m \quad (4.4.3)$$

where D_L is the hydrodynamic dispersion coefficient parallel to the main direction of flow movement (longitudinal); D_T is the hydrodynamic dispersion coefficient perpendicular to the main direction of flow movement (transverse); α_L and α_T are called longitudinal dispersion and transverse dispersion, respectively.

We cannot explain all the above observations on the basis of the average, such as with the Darcy flow. We must refer to what happens at the microscopic level, namely inside the pore. There, the velocity varies both in magnitude and direction over a given pore cross-section. We assume that the fluid velocity at the surface of the solid is zero and that the maximum velocity is at a point inside the pore (compare the parabolic velocity distribution in a straight capillary tube). The maximum velocity itself varies depending on the size of the pore. Due to the shape of the connected pore space, the (microscopic) streamlines in this space fluctuate with respect to the mean flow direction (Fig. 4.12a, b). These phenomena cause the spreading of each initially closely spaced group of tracer particles. As the flow progresses, they occupy an increasingly larger volume of the flow domain. The spreading occurs both longitudinally, in the direction of the average flow, and transversely to the average flow. We refer to this spreading, caused by the velocity variations at the microscopic level, as mechanical spreading. Sometimes it is also called convective diffusion. It is a dispersion with respect to the average flow caused by velocity variations (in magnitude and direction) in the pore space. Thus, the two fundamental elements in this type of mixing are flow and the presence of a pore system through which the flow occurs. In addition to inhomogeneity at the microscopic level (presence of pores and grains), there can also be inhomogeneity at the macroscopic level, due to variations in permeability from one part of the flow domain to the next. This inhomogeneity also contributes to the mechanical dispersion of the tracer.

In general, convective mass transport can occur both in a laminar flow, where the fluid moves along certain paths that can be averaged to obtain streamlines, and in a

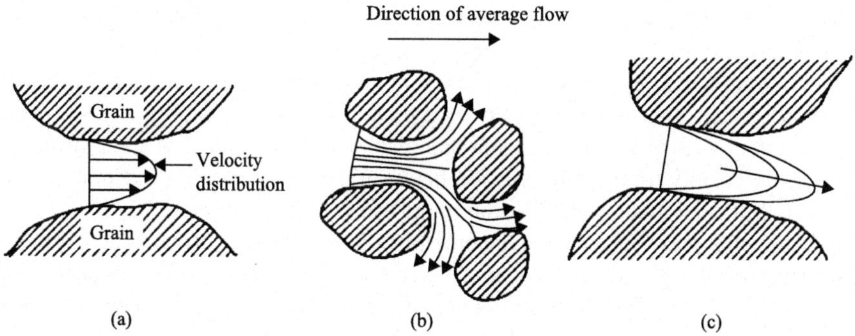

Fig. 4.12 Spreading due to mechanical dispersion and molecular diffusion

turbulent flow, where turbulence can cause additional mixing. In what follows, we will focus only on the first type of flow.

Another mass transport phenomenon that occurs simultaneously with mechanical dispersion is molecular diffusion, which is caused by fluctuations in tracer concentration in the liquid phase. Molecular diffusion causes an additional flow of tracer particles (at the microscopic level) from regions with higher tracer concentration to those with lower concentration. This means, for example, that molecular diffusion causes concentrations to equalize along the tube as the tracer spreads along the tube by mechanical dispersion, creating a concentration gradient of the tracer. At the same time, a concentration gradient also occurs between adjacent streamlines, resulting in lateral molecular diffusion through the stream tubes Fig. 4.12c.

4.5 Taylor Dispersion

Taylor dispersion is an effect in fluid mechanics in which shear flow can increase the effective diffusivity of a species. Essentially, shear causes the concentration distribution in the direction of flow to become blurred, increasing the speed of dispersion in that direction. The effect is named after the British fluid dynamicist G. I. Taylor.

Laminar, uniform flow of a viscous fluid in a cylindrical tube increases the diffusion rate of Brownian particles along the tube axis. This so-called Aris-Taylor dispersion is due to the fact that the cumulative times spent by a diffusing particle in the layers of the fluid moving at different velocities are random variables that depend on the realization of the particle's stochastic trajectory in the radial direction. A conceptually similar increase in diffusivity occurs when the particle randomly jumps between two states with different drift velocities.

Since Taylor's seminal work demonstrating that the transport of a solute in an axisymmetric shear flow can be effectively reduced to a one-dimensional dispersion process by using a longitudinal effective dispersion coefficient, the basic idea of Taylor dispersion is simple. At "asymptotic" times, i.e., times when the solute has absorbed all of the variability in flow velocities by diffusion, the solute gradients in the direction transverse to the flow direction can be considered negligible. At these times, plume spreading occurs only in the longitudinal direction and can be described by a one-dimensional advection–diffusion model. The corresponding diffusion term has an effective dispersion coefficient that includes longitudinal molecular diffusion but results mainly from the interaction between the transverse heterogeneity of the velocity field and molecular diffusion in the transverse direction.

Although it was mentioned in the previous sections that the groundwater velocity can be replaced by the average linear velocity, the velocity in the cross section is equal to 0 due to the objective constraint of the slip-free boundary condition and the shear behavior of the fluid near the crack wall, resulting in a nonuniform distribution of the velocity, as shown in Fig. 4.13. This unevenly distributed flow velocity causes the solutes to spread on the same cross-section at different velocities under the effect of convection, which increases the concentration gradient ($\partial C/\partial y$) of the solute cluster

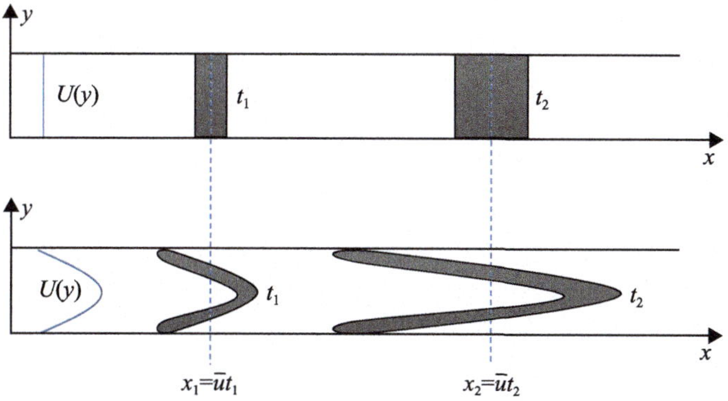

Fig. 4.13 Solute transport under slip boundary and no-slip boundary conditions in an ideal single fracture

in the transverse direction. At the same time, the molecular diffusion process under the increasing concentration gradient also accelerates the dilution of the solute cluster. Macroscopically, the migration of the solute cluster in the longitudinal direction is faster and more uneven, the concentration gradient of the solute cluster is larger in the transverse direction, and the molecular diffusion will further dilute the concentration gradient of the solute cluster in the transverse direction.

Considering the non-uniformly distributed velocity field, the solute migration process in Fig. 4.13 can be expressed as follows:

$$\left[\frac{\partial}{\partial x}\left(D_{\mathrm{m}}\frac{\partial C}{\partial x}\right)+\frac{\partial}{\partial y}\left(D_{\mathrm{m}}\frac{\partial C}{\partial y}\right)\right]-\left[\frac{\partial}{\partial x}(v_x C)+\frac{\partial}{\partial y}(v_y C)\right]=\frac{\partial C}{\partial t} \tag{4.5.1}$$

Since the velocity in the above formula is a function of spatial position, it is very difficult to obtain a simple analytical solution. If the influence of a non-uniform flow field is taken into account, this actually leads to differential solute dispersion by convection in the longitudinal direction and molecular diffusion in the transverse direction, thereby enhancing solute dilution. G. I. Taylor directly attributed the effect of a non-uniform flow field on solute migration to an additional longitudinal dispersion coefficient, also called the Taylor dispersion coefficient.

When the migration law of solute mass conforms to Fick's Law, G. I. Taylor decomposes the concentration on the cross section into the average value $\langle C\rangle$ of the concentration on the cross section and the corresponding disturbance value C', so that the average value $\langle C\rangle$ of the concentration on the cross section plus an additional longitudinal dispersion term can be directly used to show the solute migration under the influence of non-uniform distribution flow field:

$$\frac{\partial\langle C\rangle}{\partial t}+\langle u\rangle\frac{\partial\langle C\rangle}{\partial x}=D_{\mathrm{m}}\frac{\partial^2\langle C\rangle}{\partial x^2}+K_x\frac{\partial^2\langle C\rangle}{\partial x^2} \tag{4.5.2}$$

where $\langle u \rangle$ is the average value of velocity on the cross section. When $D_{\mathrm{m}} \ll K_x$, the first item on the right of the above formula can be ignored, and the above formula changes to:

$$\frac{\partial \langle C \rangle}{\partial t} + \langle u \rangle \frac{\partial \langle C \rangle}{\partial x} = K_x \frac{\partial^2 \langle C \rangle}{\partial x^2} \tag{4.5.3}$$

Assuming that the migration law of solute mass conforms to Fick's Law, the above formula has the same form as the one-dimensional convection dispersion equation and has a simple analytical solution:

$$C(x,t) = \frac{M}{A\sqrt{4\pi K_x t}} \exp\left[\frac{-(x - \langle u \rangle t)^2}{4K_x t}\right] \tag{4.5.4}$$

By fitting the measured data with the analytical solution of the above formula, the velocity K_x can be obtained. It is very important to understand the above and the tenable conditions of derivation. Therefore, considering the ideal smooth two-dimensional fracture, it is assumed that the fluid flows only in the x-direction and the flow is uniform $\partial u/\partial x = 0$, so the Eq. (4.5.1) changes as follows:

$$\frac{\partial C}{\partial t} + u\frac{\partial C}{\partial x} = \frac{\partial}{\partial x} D_{\mathrm{m}} \frac{\partial C}{\partial x} + \frac{\partial}{\partial y} D_{\mathrm{m}} \frac{\partial C}{\partial y} \tag{4.5.5}$$

The velocity u and concentration C are decomposed into the average value and the corresponding disturbance value in the y direction, where the average value is

$$\overline{u} = \frac{\int_0^b u\mathrm{d}y}{b} \tag{4.5.6}$$

$$\overline{C} = \frac{\int_0^b C\mathrm{d}y}{b} \tag{4.5.7}$$

where b is the aperture of fracture. The expressions of velocity u and concentration C after decomposition are

$$u(y) = \overline{u} + u'(y) \tag{4.5.8}$$

$$C(x, y) = \overline{C}(x) + C'(x, y) \tag{4.5.9}$$

Replace the above decomposition formula into Eq. (4.5.5), where $\partial C / \partial y = 0$, then there is

$$\frac{\partial}{\partial t}(\overline{C} + C') + (\overline{u} + u')\frac{\partial(\overline{C} + C')}{\partial x} = \frac{\partial}{\partial x} D_{\mathrm{m}} \frac{\partial(\overline{C} + C')}{\partial x} + \frac{\partial}{\partial y} D_{\mathrm{m}} \frac{\partial(C')}{\partial y} \tag{4.5.10}$$

When C' and u' are equal to 0, the above formula can be expressed as

$$\frac{\partial \overline{C}}{\partial t} + \overline{u}\frac{\partial \overline{C}}{\partial x} = \frac{\partial}{\partial x} D_{\mathrm{m}} \frac{\partial (\overline{C})}{\partial x} + \overline{u' \frac{\partial C'}{\partial x}} \tag{4.5.11}$$

From the above formula, it can be seen that the concentration is only a function in the x-direction. A new term appears in the second term on the right side of the above formula, representing the flux associated with the velocity and concentration perturbation values. Since Eqs. (4.5.10) and (4.5.11) must be set up simultaneously, Eq. (4.5.11) must be replaced in Eq. (4.5.10):

$$\frac{\partial C'}{\partial t} + u'\frac{\partial \overline{C}}{\partial x} + u'\frac{\partial C'}{\partial x} - \overline{u' \frac{\partial C'}{\partial x}} = \frac{\partial}{\partial y} D_{\mathrm{m}} \frac{\partial (C')}{\partial y} \tag{4.5.12}$$

When $C' \ll \overline{C}$, in order to make Eqs. (4.5.10) and (4.5.11) hold at the same time, the third and fourth terms on the left of Eq. (4.5.12) must be much smaller than the second term on the left. Therefore, after ignoring these smaller terms, the equation becomes:

$$\frac{\partial C'}{\partial t} + u'\frac{\partial \overline{C}}{\partial x} = \frac{\partial}{\partial y} D_{\mathrm{m}} \frac{\partial (C')}{\partial y} \tag{4.5.13}$$

The above formula actually describes that differential convection in the x-direction and molecular diffusion in the y-direction are the reasons for the disturbance of concentration deviation from its average value. When the solute begins to migrate under the action of non-uniform flow field, the $\partial \overline{C}/\partial x$ dominated by differential convection $\partial \overline{C}/\partial x$ value is very large, and relative to the molecular diffusion in y-direction, the perturbation value of concentration deviation from its average value is larger. When the solute continues to migrate under the action of non-uniform flow field, $\partial \overline{C}/\partial x$ value will continue to decrease and eventually become very small.

At this time, the disturbance value of concentration deviation from its average value caused by differential convection is less than that caused by molecular diffusion in y-direction. When differential convection and molecular diffusion in the y-direction have the same effect on solute migration, the $\partial C'/\partial t$ tends to 0. Equation (4.5.13) is transformed into:

$$u'\frac{\partial \overline{C}}{\partial x} = \frac{\partial}{\partial y} D_{\mathrm{m}} \frac{\partial C'}{\partial y} \tag{4.5.14}$$

When the boundary condition is

$$\frac{\partial C'}{\partial y} = 0, \quad y = 0 \text{ and } b \tag{4.5.15}$$

The corresponding analytical solution is

$$C'(y) = \frac{\partial \overline{C}}{\partial x} \int_0^b \frac{1}{D_{\mathrm{m}}} \int_0^b u' \mathrm{d}y \mathrm{d}y + C'(0) \tag{4.5.16}$$

By $\int_0^b u'C'(0)\mathrm{d}y = 0$, the right-hand term in Eq. (4.5.13) can be solved using the above analytical solution, and then the time-dependent dispersion coefficient can be determined. If the migration time of the solute is long enough, the theoretical formula of the Taylor dispersion coefficient for an ideal smooth two-dimensional fracture is as follows:

$$D_{\mathrm{taylor}} = D_{\mathrm{m}} + \frac{(\overline{u}\overline{b})^2}{210 D_{\mathrm{m}}} \tag{4.5.17}$$

It should be noted that the above formula is valid if the solute migration time is long enough, that is, the gradual time is reached. In order to explain the change of dispersion coefficient related to time, the spatial moment is used to calculate the dispersion process of solute clusters. The molecular diffusion coefficient is assumed to be $D_{\mathrm{m}} = 1 \times 10^{-9}$ m/s.

Taking the ideal smooth two-dimensional crack as an example, the length of the ideal smooth two-dimensional crack is $L = 200$ mm, the gap width is $b = 0.5$ mm, the ratio of length to gap width is $L/b = 400$, and the corresponding asymptotic time is $\tau = b^2/4D_{\mathrm{m}} = 62.5$ s. The average flow velocity in the fracture is $\overline{u} = 6.17 \times 10^{-5}$ m/s.

The total mass of solute in the calculation area is calculated by using the spatial moment of solute cluster

$$M = \int_{\Omega} C(x, t) \mathrm{d}\Omega \tag{4.5.18}$$

First moment in x-direction:

$$M_{1,x}(t) = \int_{\Omega} C(x, t) x \mathrm{d}\Omega \tag{4.5.19}$$

First moment in y-direction:

$$M_{1,y}(t) = \int_{\Omega} C(x, t) y \mathrm{d}\Omega \tag{4.5.20}$$

Second moment in x-direction:

$$M_{2,xx}(t) = \frac{\int_{\Omega} C(x, t)(x - M_{1,x}/M)(x - M_{1,x}/M) \mathrm{d}\Omega}{M} \tag{4.5.21}$$

Second moment in y-direction:

$$M_{2,yy}(t) = \frac{\int_\Omega C(x,t)(x - M_{1,y}/M)(x - M_{1,y}/M)\mathrm{d}\Omega}{M} \tag{4.5.22}$$

Second order moment in x–y diagonal direction:

$$M_{2,xy}(t) = \frac{\int_\Omega C(x,t)(x - M_{1,x}/M)(x - M_{1,y}/M)\mathrm{d}\Omega}{M} \tag{4.5.23}$$

As shown in Fig. 4.14, before the progressive time, the dispersion coefficient is not a fixed value, but a function related to time. According to the physical meaning of spatial moment, the dispersion coefficient in x-direction can be expressed by the change rate of second-order moment in x-direction,

$$2D_x = \lim_{t\to\infty} \frac{\mathrm{d}M_{2,xx}}{\mathrm{d}t} \tag{4.5.24}$$

From Fig. 4.14, we can see that the dispersion coefficient increases with time before progressive time; near asymptotic time, the dispersion coefficient calculated by the second-order moment in the x-direction is close to the theoretical value of the Taylor dispersion coefficient, but it is still smaller than the theoretical value of the Taylor dispersion coefficient. This is because we use only the second moment in the x-direction to calculate the total dispersion coefficient, while ignoring the changes in the solute clusters in other directions. Furthermore, when deriving the Taylor dispersion coefficient, the influence of molecular diffusion in the x-direction is ignored if the solute migration time exceeds the asymptotic time.

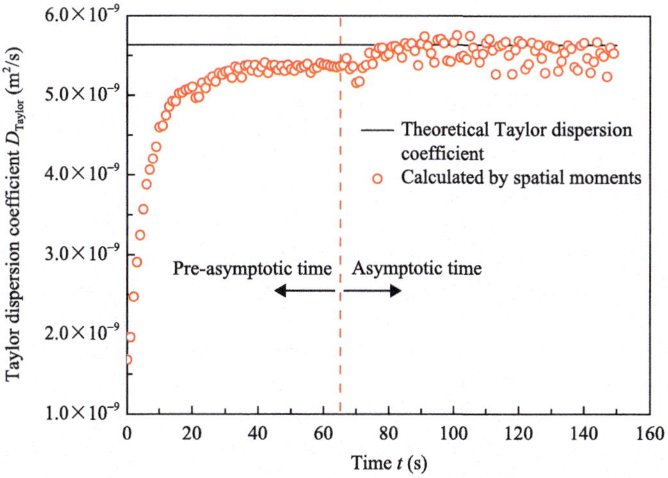

Fig. 4.14 Dynamic variation process of Taylor dispersion coefficient

4.6 Adsorption and Desorption

Mass transfer between mobile groundwater and minerals occurs through a thin layer of physically bound water, the Nernst layer. In addition to the forces of thermal (Brownian) motion, interatomic and intermolecular interactions, which are also involved in homogeneous processes, the electrostatic forces of the charged surface of the solid act. Their effect increases as the interface is approached and is manifested in the differentiation of ions according to the sign and value of their charge. Within the Nernst layer, the concentrations of ions of opposite sign to the sign of the surface increase and of the same sign decrease. Such changes in water composition are the essence of adsorption. This process is related to the ability of ions to overcome the slip plane, i.e. a reversible process called ion exchange. In contrast, the interaction forces directly at the interface can be so strong that coordination bonds are formed with atoms of the crystal lattice and the solute ion becomes immobile as part of the solid. This process is called chemisorption. Its consequences are the formation of surface complexes, formation of minerals, deposits, etc. The relaxation of mass transfer between mineral and groundwater takes place under the conditions of the electrostatic field of the mineral surface. For this reason, within the Nernst layer is formed the solution of a special state, the structure and composition of which depends on the distance from the interface.

4.7 Precipitation and Dissolution

Precipitation and dissolution are good examples of dynamic equilibrium (also described here). Whenever there is a solution with a small amount of solid in it, both processes occur simultaneously. Some molecules or ions leave the solid and become solvated, and some solvated dissolved particles collide with the solid and remain suspended. The speed of the two processes determines the overall effect: if precipitation proceeds faster, a great deal of solid can leave the solution very quickly. If dissolution proceeds faster, the solids dissolve. As the concentration of the solution increases, the rate of precipitation increases and the rate of dissolution decreases, so that eventually the concentration no longer changes. When equilibrium is reached, the solution is saturated, and this concentration defines the solubility of the solute. Solubility is the maximum concentration possible and is expressed in M, g/L, or other units. Solubility changes with temperature. So when you look up data on solubility, the temperature is given.

Precipitation is the process by which a compound leaves solution. It is the opposite of dissolution or solvation. In dissolution, the solute particles separate from each other and become surrounded by solvent molecules. In precipitation, the solute particles find each other and together form a solid. This solid is called a precipitate or sometimes abbreviated as "ppt".

Precipitation can occur for a variety of reasons, such as when a solution has been cooled or some of the solvent has been removed by evaporation, or both (this is often used to purify a compound). Precipitation can also occur when two solutions are mixed and a new combination of ions becomes supersaturated in the combined solution. For example, you have mixed a solution of silver(I) nitrate and sodium chloride. Since silver(I) chloride is very insoluble, it will precipitate leaving the soluble sodium nitrate in the solution. Precipitation reactions can be a good way to make a desired salt from other salts with the correct anion and cation. Precipitation reactions can also be used to detect the presence of certain ions in a solution. For example, chloride, iodide, and bromide can be detected in an unknown solution by adding silver(I) ions and watching for precipitation.

Figure 4.15a shows that when a solid is added to a solvent in which it is soluble, the solute particles leave the surface of the solid and are dissolved by the solvent, initially forming an unsaturated solution. Figure 4.15b shows that the solution is saturated when the largest possible amount of solute has dissolved. When there is an excess of solute, the rate at which the solute leaves the surface of the solid is equal to the rate at which it returns to the surface of the solid. Figure 4.15c shows that a supersaturated solution can usually be formed from a saturated solution by filtering off the excess solute and lowering the temperature. Figure 4.15d shows that when a solute seed crystal is added to a supersaturated solution, the solute particles leave the solution and form a crystalline precipitate.

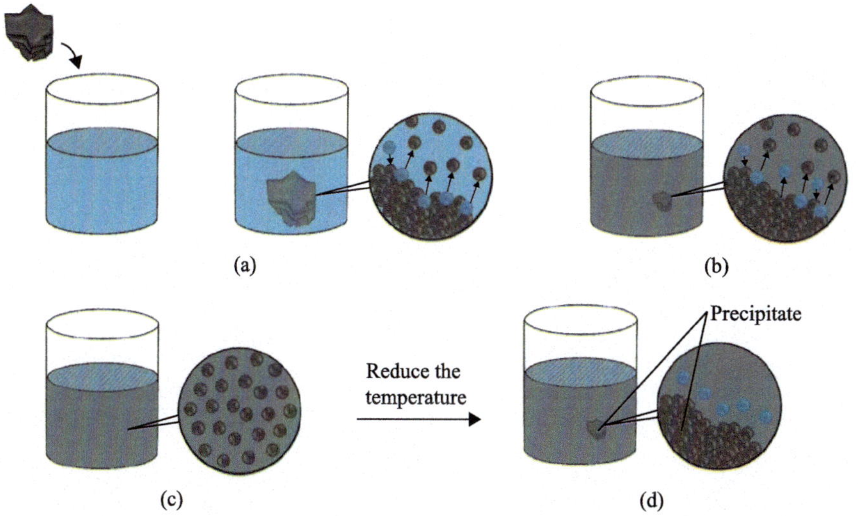

Fig. 4.15 Precipitation processes **a** unsaturated solution, **b** saturated solution, **c** supersaturated solution, **d** precipitate forms

Chapter 5
Mathematical Model of Mass Transfer in Fractured Media

To model fluid dynamics, mass transfer, and other relevant physical phenomena, it is essential to mathematically express the underlying physics. The conservation principles govern nearly all the physical phenomena of interest discussed in this article, and they are described by partial differential equations that reflect these principles. Specifically, the momentum equations reflect the conservation of linear momentum, while the energy equation expresses the conservation of total energy. This chapter focuses on the derivation of a typical conservation equation and its mathematical properties.

5.1 Analytical Solution of Advection–Dispersion Equation (ADE) Model

5.1.1 ADE Model and Analytical Solution in One-Dimensional Fractured Media

Although the validity of the one-dimensional (1D) Advection–Dispersion Equation (ADE) model has been confirmed only for homogeneous geological formations in capturing Fickian transport behavior, we employed it as a reference point for comparison with an advanced model, specifically the Continuous-Time Random Walk (CTRW) Two-Phase Liquid (TPL) model. The ADE model is based on Fick's law, which states that the dispersive mass flux is proportional to the first-order spatial derivative of the concentration. The governing equation for the 1D ADE model is expressed as follows:

$$\frac{\partial C}{\partial t} + \overline{u}\frac{\partial C}{\partial x} = D_{\text{ADE}}\frac{\partial^2 C}{\partial x^2} \tag{5.1.1}$$

Z. Dou et al., *Mass Transfer Dynamics of Contaminants in Fractured Media*,
https://doi.org/10.1007/978-981-99-9187-7_5

where D_{ADE} is the effective dispersion and $\overline{u}$ is the average flow velocity. The corresponding analytical solution is given by

$$C(x,t)=\frac{C_0}{2}\left[\operatorname{erfc}\left(\frac{L-\overline{u}t}{2\sqrt{D_{\mathrm{ADE}}t}}\right)+\exp\left(\frac{\overline{u}L}{D_{\mathrm{ADE}}}\right)\operatorname{erfc}\left(\frac{L+\overline{u}t}{2\sqrt{D_{\mathrm{ADE}}t}}\right)\right] \tag{5.1.2}$$

where erfc is the complementary error function. The D_{ADE} can be obtained by fitting the flux-weight breakthrough curves (BTCs).

5.1.2 ADE Model and Analytical Solution in Two-Dimensional Fractured Media

It is assumed that the Advection–Dispersion Equation (ADE) sufficiently characterizes the solute transport. The one-dimensional advection term and two-dimensional dispersion term govern the transport phenomena. The steady flow conditions in a homogeneous and isotropic medium are represented as follows:

$$\frac{\partial C}{\partial t}=D_{\mathrm{L}}\frac{\partial^2 C}{\partial x^2}-v\frac{\partial C}{\partial x}+D_{\mathrm{T}}\frac{\partial^2 C}{\partial y^2} \tag{5.1.3}$$

where C is the solute concentration [ML^{-3}], t is time [T], D_{L} and D_{T} are the coefficients of longitudinal and transverse dispersion [$\mathrm{L}^2\mathrm{T}^{-1}$], respectively, v is the pore water velocity [LT^{-1}], and x and y are positions at the coordinate axes parallel and perpendicular to the direction of flow [L]. The solution domain is a half plane with $x \geq 0$ and the other boundaries at infinity. The problem is solved for a prescribed concentration at $x = 0$ (a Dirichlet condition). The initial and boundary conditions are (see also the section on the effect of transverse dispersion on transport):

$$\left.\frac{\partial C}{\partial x}\right|_{x\to\infty}=0,\quad -\infty<y<+\infty,\quad t\geq 0 \tag{5.1.4}$$

$$C(0,y,t)=g(y)=\begin{cases}C_{\mathrm{L}}, & y<0\\ C_{\mathrm{R}}, & y>0\end{cases},\ t\geq 0 \tag{5.1.5}$$

$$C(x,y,0)=f(x,y),\quad 0<x<\infty,\quad -\infty<y<+\infty \tag{5.1.6}$$

$$\left.\frac{\partial C}{\partial x}\right|_{y\to\pm\infty}=0,\quad 0<x<\infty \tag{5.1.7}$$

These are also the experimental conditions for the determination of D_{T}. Figure 5.1 illustrates the problem to be solved.

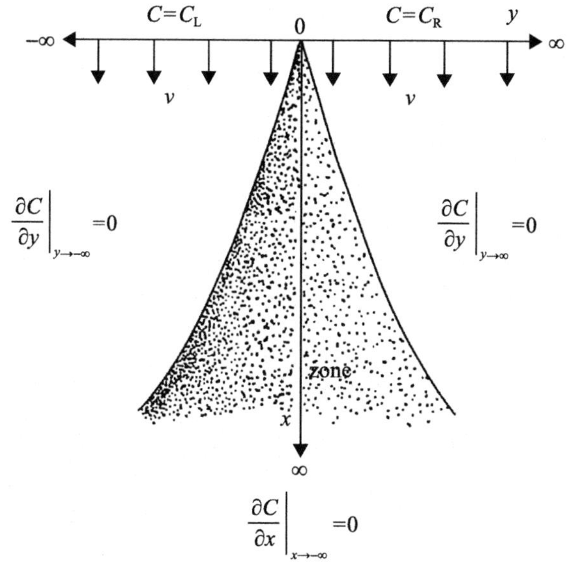

Fig. 5.1 Schematic of 1D advection and 2D dispersion in a half plane for an isotropic medium

Previous studies have addressed solutions for the problem under steady-state conditions. Grane and Gardner utilized a transformation to parabolic coordinates, resulting in a diffusion equation. However, a disadvantage of this method is that the transformation leads to complex boundary conditions, which is a well-known limitation of using a moving coordinate system in solving the one-dimensional Advection–Dispersion Equation (ADE). Other scholars have proposed analytical solutions for a semi-infinite system, wherein longitudinal dispersion is disregarded $D_L \frac{\partial^2 C}{\partial x^2} \ll D_T \frac{\partial^2 C}{\partial y^2}$. However, the transient-state solution is usually of interest for contaminant transport. It is therefore desirable to obtain asolution, which facilitates the simultaneous determination of D_L and D_T.

Certain researchers employed the separation of variables method to obtain an infinite series solution for a finite system. However, this approach can be rather intricate, and since we are essentially dealing with a half-plane, an alternative solution technique was employed.

The solution to Eq. (5.1.3), subject to Eqs. (5.1.4)–(5.1.7), can be obtained using a double integral transform technique, specifically a Laplace transform with respect to t and a Fourier transform with respect to y. This approach combines solution techniques for both the one-dimensional Advection–Dispersion Equation (ADE) and the two-dimensional ADE for steady-state conditions. It is worth mentioning that the diffusion equation for a half-plane can be solved in a comparable manner, with the exception that a Laplace transform is taken with respect to x.

First, the Laplace transform is taken. For completeness, the Laplace transforms of C and $\partial C/\partial t$ are given:

$$[C(x, y, t)] = \int_0^{\infty} C(x, y, t) \exp(-pt) \mathrm{d}t = \overline{C}(x, y, p) \tag{5.1.8}$$

$$\varphi\left[\frac{\partial C}{\partial t}\right] = p\overline{C} - C(x, y, 0) \tag{5.1.9}$$

where p is the fluid pressure.

The Laplace transforms of Eqs. (5.1.3) and (5.1.4)–(5.1.7) are:

$$p\overline{C} - f = D_{\mathrm{L}} \frac{\partial^2 \overline{C}}{\partial x^2} - v \frac{\partial \overline{C}}{\partial x} + D_{\mathrm{T}} \frac{\partial^2 \overline{C}}{\partial y^2} \tag{5.1.10}$$

$$\left. \frac{\partial \overline{C}}{\partial x} \right|_{x \to \infty} = 0 \tag{5.1.11}$$

$$\overline{C}(0, y, p) = \frac{g}{p} \tag{5.1.12}$$

Proceed by taking the Fourler transform for the infinite y-domain:

$$\mathcal{F}\left[\overline{C}(x, y, p)\right] = \frac{1}{\sqrt{2\pi}} \int_{-\infty}^{+\infty} \overline{C}(x, y, p) \exp(i\alpha y) \mathrm{d}y = \tilde{\overline{C}}(x, \alpha, p) \tag{5.1.13}$$

$$\mathcal{F}\left[\frac{\partial^2 \overline{C}}{\partial y^2}\right] = -\alpha^2 \tilde{\overline{C}} \tag{5.1.14}$$

The Fourier transforms of Eqs. (5.1.10) and (5.1.11), (5.1.12) are

$$D_{\mathrm{L}} \frac{\mathrm{d}^2 \overline{C}}{\mathrm{d}x^2} - v \frac{\mathrm{d}\overline{C}}{\mathrm{d}x} - (\alpha^2 D_{\mathrm{T}} + p)\tilde{\overline{C}} + \tilde{f} = 0 \tag{5.1.15}$$

$$\left. \frac{\mathrm{d}\overline{C}}{\mathrm{d}x} \right|_{x \to \infty} = 0 \tag{5.1.16}$$

$$\tilde{\overline{C}}(0, \alpha, p) = \frac{\tilde{g}}{p} \tag{5.1.17}$$

where $\tilde{f}$ and $\tilde{g}$ are the Fourier transforms of f and g.

A solution of Eq. (5.1.15), an ordinary differential equation, subject to Eqs. (5.1.16) and (5.1.17) is:

$$\tilde{\overline{C}}(x, \alpha, p) = A \exp(\lambda_1 x) + B \exp(\lambda_2 x) + \frac{1}{D_T \alpha^2 + p} \tag{5.1.18}$$

where A and B are constants determined by the boundary conditions and where the roots $\lambda_{1,2}$ are given by:

$$\lambda_{1,2} = \frac{v}{2D_L} \pm \sqrt{\left(\frac{v}{2D_L}\right)^2 + \frac{D_T}{D_L}\alpha^2 + \frac{p}{D_L}} \tag{5.1.19}$$

It follows from Eq. (5.1.18) that $A = 0$. Using the boundary condition at the inlet, Eq. (5.1.17), we obtain:

$$B = \frac{\tilde{g}}{p} - \frac{\tilde{f}}{D_T\alpha^2 + p} \tag{5.1.20}$$

Substitution of Eq. (5.1.20) into Eq. (5.1.18) yields:

$$\tilde{C}(x, \alpha, p) = \left(\frac{\tilde{g}}{p} - \frac{\tilde{f}}{D_T\alpha^2 + p}\right)\exp(\lambda_2 x) + \frac{\tilde{f}}{D_T\alpha^2 + p} \tag{5.1.21}$$

For a solution in the x, y, t-domain, inverse transformations need to be carried out. Although numerical inversion technlques are available, the use of analytlcal methods is preferred. First, the inverse Laplace transformation is carried out, for which purpose the right-hand side of Eq. (5.1.21) is split up into three terms, the first one being:

$$\begin{aligned}\overline{C}_1(x, a, t) &= \varphi^{-1}\left[\tilde{C}_1(x, \alpha, p)\right] = \varphi^{-1}\left[\frac{\tilde{g}}{p}\exp(\lambda_2 x)\right] \\ &= \tilde{g}\exp\left(\frac{vx}{2D_L}\right)\varphi^{-1}\left[\frac{1}{p}\exp\left(-\frac{x}{\sqrt{D_L}}\sqrt{\frac{v^2}{4D_L} + D_T\alpha^2 + p}\right)\right]\end{aligned} \tag{5.1.22}$$

The inverse Laplace operator is evaluated with the help of the convolution theorem. For the functions $h(t)$ and $k(t)$, with Laplace convolution theorem. For the functions $h(t)$ and $k(t)$, with Laplace transforms $\overline{h}(p)$ and $\overline{k}(p)$, the following convolution integrals can be defined:

$$\varphi^{-1}\left[\overline{h}(p)\,\overline{k}(p)\right] = h^*k = \int_0^t h(t-\xi)k(\xi)\mathrm{d}\xi = \int_0^t h(\xi)\,k(t-\xi)\mathrm{d}\xi \tag{5.1.23}$$

where ξ is a dummy variable. Two functions can be distinguished in the expression of Eq. (5.1.22), which need to be inverted:

$$\overline{h}(p) = \frac{1}{p} \tag{5.1.24}$$

$$\overline{k}(p)=\exp\left(-\frac{x}{\sqrt{D_{\mathrm{L}}}}\sqrt{\frac{v^2}{4D_{\mathrm{L}}}+D_{\mathrm{T}}\alpha^2+p}\right) \tag{5.1.25}$$

The inverse expressions, $h(t)$ and $k(t)$, are determined with the help of the shift theorem and a table of Laplace transforms:

$$h(t)=1 \tag{5.1.26}$$

$$k(t)=\frac{x}{t\sqrt{4\pi D_{\mathrm{L}}t}}\exp\left[-\left(\frac{v^2}{4D_{\mathrm{L}}}+D_{\mathrm{T}}\alpha^2\right)t-\frac{x^2}{4D_{\mathrm{L}}t}\right] \tag{5.1.27}$$

Substitution in Eq. (5.1.22), results in

$$\tilde{C}_1(x,\alpha,t)=\tilde{g}\exp\left(\frac{vx}{2D_{\mathrm{L}}}\right)\int_0^t\frac{x}{\xi\sqrt{4\pi D_{\mathrm{L}}\xi}}\exp\left[-\left(\frac{v^2\xi^2+4D_{\mathrm{L}}D_{\mathrm{T}}\alpha^2\xi^2+x^2}{4D_{\mathrm{L}}\xi}\right)\right]\mathrm{d}\xi \tag{5.1.28}$$

The second term, $\tilde{\overline{C}}_2(x,\alpha,p)$, is inverted in a similar manner, except that no convolution integral is needed.

$$\begin{aligned}\tilde{C}_2(x,\alpha,t)&=\varphi^{-1}\left[\tilde{\overline{C}}_2(x,\alpha,p)\right]=\varphi^{-1}\left[-\frac{\tilde{f}}{D_{\mathrm{T}}\alpha^2+p}\exp(\lambda_2x)\right]\\&=-\overline{f}\exp\left(\frac{vx}{2D_{\mathrm{L}}}\right)\exp\left[-\left(\frac{v^2}{4D_{\mathrm{L}}}+D_{\mathrm{T}}\alpha^2\right)t\right]\varphi^{-1}\frac{1}{p-\frac{v^2}{4D_{\mathrm{L}}}}\\&=\frac{-\overline{f}}{2}\exp(-D_{\mathrm{T}}\alpha^2t)\left[\operatorname{erfc}\left(\frac{x-vt}{\sqrt{4D_{\mathrm{L}}t}}\right)+\exp\left(\frac{vx}{D_{\mathrm{L}}}\right)\operatorname{erfc}\left(\frac{x+vt}{\sqrt{4D_{\mathrm{L}}t}}\right)\right]\end{aligned} \tag{5.1.29}$$

where erfc is the complementary error function. The third term of Eq. (5.1.21) can be directly obtained from a table of Laplace transforms:

$$\tilde{C}_3(x,\alpha,t)=\varphi^{-1}\left[\tilde{\overline{C}}_3(x,\alpha,p)\right]=\varphi^{-1}\left(\frac{\tilde{f}}{D_{\mathrm{T}}\alpha^2+p}\right)=\tilde{f}\exp(-D_{\mathrm{T}}\alpha^2t) \tag{5.1.30}$$

The resulting expression for $\tilde{C}(x,\alpha,t)$, the Fourier transform of the analytical solution being sought, is:

$$\tilde{C}(x,\alpha,t)=\tilde{C}_1+\tilde{C}_2+\tilde{C}_3 \tag{5.1.31}$$

The last step of the solution procedure is the application of the inverse Fourier transform to each of the terms in Eq. (5.1.31). The inverse transformation of $\tilde{C}(x, \alpha, t)$ is formally given by:

$$C(x, y, t) = \mathcal{F}^{-1}\left[\tilde{C}(x, \alpha, t)\right] = \frac{1}{\sqrt{2\pi}} \int_{-\infty}^{\infty} \tilde{C} \exp(-i\alpha y) d\alpha \tag{5.1.32}$$

The inverse of the first term of Eq. (5.1.31) is determined as follows.

$$\begin{aligned} C_1(x, y, t) &= \mathcal{F}^{-1}\left[\tilde{C}_1(x, \alpha, t)\right] \\ &= \frac{1}{\sqrt{2\pi}} \int_{-\infty}^{\infty} \exp\left(\frac{vx}{2D_L}\right) \int_0^t \frac{\tilde{g}x}{\xi\sqrt{4\pi D_L \xi}} \exp\left[-\left(\frac{v^2\xi^2 + 4D_L D_T \alpha^2 \xi^2 + x^2}{4D_L \xi}\right)\right] d\xi \exp(-i\alpha y) d\alpha \\ &= \int_0^t \frac{1}{\sqrt{2\pi}} \left\{ \int_{-\infty}^{\infty} \frac{\tilde{g}x}{\xi\sqrt{4\pi D_L \xi}} \exp\left(\frac{vx}{2D_L}\right) \exp\left[-\left(\frac{v^2\xi^2 + 4D_L D_T \alpha^2 \xi^2 + x^2}{4D_L \xi}\right)\right] \exp(-i\alpha y) d\alpha \right\} d\xi \end{aligned} \tag{5.1.33}$$

Again, the convolution $h * k$ was used to determine the inverse of a product of two functions $\tilde{h}$ and $\tilde{k}$, which are now in the Fourier domain:

$$\frac{1}{\sqrt{2\pi}} \int_{-\infty}^{\infty} \tilde{h}(\alpha)\tilde{k}(\alpha) \exp(-i\alpha y) d\alpha = h * k = \frac{1}{\sqrt{2\pi}} \int_{-\infty}^{\infty} h\,(y) k(y - v) dv \tag{5.1.34}$$

After inspection of the last expression for $C_1(x, y, t)$ in Eq. (5.1.33), the following two functions seem convenient for the convolution integral:

$$\tilde{h}(\alpha) = \tilde{g} \tag{5.1.35}$$

$$\tilde{k}(\alpha) = \frac{x}{\xi\sqrt{4\pi D_L \xi}} \exp\left[-\left(\frac{x - v\xi}{\sqrt{4D_L \xi}}\right)^2\right] \exp(-D_T \alpha^2 \xi) \tag{5.1.36}$$

Without actually determining $\tilde{g}$, which is the Fourier transform of a step function, it is clear that $h(y)$ must equal $g(y)$. To determine the inverse of $\tilde{k}(\alpha)$, note that:

$$\mathcal{F}^{-1}\left[\exp(-D_T \alpha^2 \xi)\right] = \frac{1}{\sqrt{2\xi D_T}} \exp\left(-\frac{y^2}{4\xi D_T}\right) \tag{5.1.37}$$

The resulting $h(y)$ and $k(y)$ are:

$$h(y) = g(y) = \begin{cases} C_L, & y < 0 \\ C_R, & y > 0 \end{cases} \tag{5.1.38}$$

$$k(y) = \frac{1}{\sqrt{2\pi}} \int_{-\infty}^{\infty} \tilde{k}(\alpha) \exp(-i\alpha y) \mathrm{d}\alpha$$
$$= \frac{x}{\xi \sqrt{4\pi D_\mathrm{L} \xi}} \exp\left[\left(\frac{x - v\xi}{\sqrt{4D_\mathrm{L}}}\right)^2\right] \frac{1}{\sqrt{2\xi D_\mathrm{T}}} \exp\left(-\frac{y^2}{4\xi D_\mathrm{T}}\right) \tag{5.1.39}$$

From Eq. (5.1.34) it follows that:

$$h * k = \frac{1}{\sqrt{2\pi}} \int_{-\infty}^{\infty} \frac{xh(v)}{\xi \sqrt{4\pi D_\mathrm{L} \xi}} \exp\left[-\left(\frac{x - v\xi}{\sqrt{4D_\mathrm{L}}}\right)^2\right] \frac{1}{\sqrt{2\xi D_\mathrm{T}}} \exp\left[-\frac{(y - v)^2}{4\xi D_\mathrm{T}}\right] \mathrm{d}v \tag{5.1.40}$$

This expression is evaluated by using the substitution $\rho = \frac{v-y}{2\sqrt{\xi D_\mathrm{T}}}$, Eq. (5.1.38) and the properties of the error function:

$$h * k$$
$$= \frac{x}{\xi \sqrt{4\pi D_\mathrm{L} \xi}} \exp\left[-\left(\frac{x - v\xi}{\sqrt{4D_\mathrm{L}}}\right)^2\right] \left[\frac{C_\mathrm{L}}{2} \mathrm{erfc}\left(\frac{y}{\sqrt{4\xi D_\mathrm{T}}}\right) + \frac{C_\mathrm{R}}{2} \mathrm{erfc}\left(-\frac{y}{\sqrt{4\xi D_\mathrm{T}}}\right)\right] \tag{5.1.41}$$

This result can be substituted in Eq. (5.1.33) to obtain the following expression for $C_1(x, y, t)$:

$$C_1(x, y, t) = \int_0^t h * k d\xi$$
$$= \int_0^t \frac{x}{\xi \sqrt{4\pi D_\mathrm{L} \xi}} \exp\left[-\left(\frac{x - v\xi}{\sqrt{4D_\mathrm{L} \xi}}\right)^2\right] \left[\begin{array}{l} \frac{C_\mathrm{L}}{2} \mathrm{erfc}\left(\frac{y}{\sqrt{4\xi D_\mathrm{T}}}\right) \\ + \frac{C_\mathrm{R}}{2} \mathrm{erfc}\left(-\frac{y}{\sqrt{4\xi D_\mathrm{T}}}\right) \end{array}\right] \mathrm{d}\xi \tag{5.1.42}$$

The second term, $C_2(x, y, t)$, is given as follows:

$$C_2(x, y, t) = -\frac{1}{2}\left[\mathrm{erfc}\left(\frac{x - vt}{2\sqrt{D_\mathrm{L} t}}\right) + \exp\left(\frac{vx}{D_\mathrm{L}}\right) \mathrm{erfc}\left(\frac{x + vt}{2\sqrt{D_\mathrm{L} t}}\right)\right]$$
$$\mathcal{F}^{-1}\left[\tilde{f} \exp(-D_\mathrm{T} \alpha^2 t)\right] \tag{5.1.43}$$

The Fourier inverse term in Eq. (5.1.43) is obtained with the convolution theorem, with:

$$\tilde{h}(\alpha) = \tilde{f} \tag{5.1.44}$$

$$\tilde{k}(\alpha) = \exp(-D_{\mathrm{T}}\alpha^2 t) \tag{5.1.45}$$

For convenience, it is assumed that the initial concentration is constant, viz. C_{i}. Making use of Eq. (5.1.37) results in the following inverse functions:

$$h(y) = f(x, y) = C_{\mathrm{i}} \tag{5.1.46}$$

$$k(y) = \frac{1}{\sqrt{2D_{\mathrm{T}}t}}\exp\left(-\frac{y^2}{4D_{\mathrm{T}}t}\right) \tag{5.1.47}$$

The inverse transformation can now be carried out by using the properties of erfc and applying the same substitution variable ρ:

$$\mathcal{F}^{-1}[\tilde{f}\exp(-D_{\mathrm{T}}\alpha^2 t)] = h * k = \frac{C_{\mathrm{i}}}{\sqrt{4\pi D_{\mathrm{T}}t}}\int_{-\infty}^{\infty}\exp\left[-\frac{(y-v)^2}{4D_{\mathrm{T}}t}\right]\mathrm{d}v = C_{\mathrm{i}} \tag{5.1.48}$$

Hence:

$$C_2(x, y, t) = -\frac{C_{\mathrm{i}}}{2}\left[\mathrm{erfc}\left(\frac{x - vt}{\sqrt{4D_{\mathrm{L}}t}}\right) + \exp\left(\frac{vx}{D_{\mathrm{L}}}\right)\mathrm{erfc}\left(\frac{x + vt}{\sqrt{4D_{\mathrm{L}}t}}\right)\right] \tag{5.1.49}$$

The last term, $C_3(x, y, t)$, was already evaluated in Eq. (5.1.49):

$$C_3(x, y, t) = \mathcal{F}^{-1}\left[\tilde{f}\exp(-D_{\mathrm{T}}\alpha^2 t)\right] = C_{\mathrm{i}} \tag{5.1.50}$$

The resulting expression for $C(x, y, t)$ is:

$$\begin{aligned} C(x, y, t) = {} & \int_0^t \frac{x}{\xi\sqrt{4\pi D_{\mathrm{L}}\xi}}\left[\frac{C_{\mathrm{L}}}{2}\mathrm{erfc}\left(\frac{y}{2\sqrt{D_{\mathrm{T}}\xi}}\right) + \frac{C_{\mathrm{R}}}{2}\mathrm{erfc}\left(-\frac{y}{2\sqrt{D_{\mathrm{T}}\xi'}}\right)\right] \\ & \times \exp\left[-\left(\frac{x - v\xi}{2\sqrt{D_{\mathrm{L}}}}\right)^2\right]\mathrm{d}\xi - \frac{C_{\mathrm{i}}}{2}\left[\mathrm{erfc}\left(\frac{x - vt}{2\sqrt{D_{\mathrm{L}}t}}\right) + \exp\left(\frac{vx}{D_{\mathrm{L}}}\right)\mathrm{erfc}\left(\frac{x + vt}{2\sqrt{D_{\mathrm{L}}t}}\right)\right] + C_{\mathrm{i}} \end{aligned} \tag{5.1.51}$$

Numerical evaluation of the integral in Eq. (5.1.51) can be accomplished using the Gauss–Chebyshev quadrature, which is a convenient method. A computer program can be employed to calculate the integral, which includes the function EXF(*A*,*B*) to compute exp(*A*)erfc(*B*). It is important to note that the solution remains sufficiently general by avoiding any specification of f and g. In many other instances, a simpler expression arises if C_{L} or C_{R} and C_{i} are equal to 0. Transport of solutes which are linearly retarded, because of reactions with the solid phase of the medium, can be described by dividing the parameters D_L, D_T, and v with the retardation coefficient.

5.1.3 ADE Model and Analytical Solution in Three-Dimensional Fractured Media

Figure 5.2 provides a schematic representation of the geometry associated with the problem addressed by the Domenico and Sagar solutions. The equation that characterizes advective–dispersive transport within a uniform flow field, featuring linear retardation and first-order decay, can be expressed as:

$$R\frac{\partial C}{\partial t} = D_x\frac{\partial^2 C}{\partial x^2} + D_y\frac{\partial^2 C}{\partial y^2} + D_z\frac{\partial^2 C}{\partial z^2} - vC - R\lambda C \tag{5.1.52}$$

with:

$$D_x = \alpha_x v + D_0\psi, \quad D_y = \alpha_y v + D_0\psi, \quad D_z = \alpha_z v + D_0\tau \tag{5.1.53}$$

where C is the solute concentration at points x, y, z and at time t; v is the average linear groundwater velocity; α_x is longitudinal dispersivity (in the direction of flow); α_y is horizontal-transverse dispersivity; α_z is vertical-transverse dispersivity; D_0 is free solution diffusion coefficient; τ is tortuosity; R is retardation coefficient in the aquifer; and λ is the first-order constant for decay in the aquifer. Embedded in Eq. (5.1.52) are the following assumptions:

(1) Groundwater flow is steady and unidirectional along the x-axis.

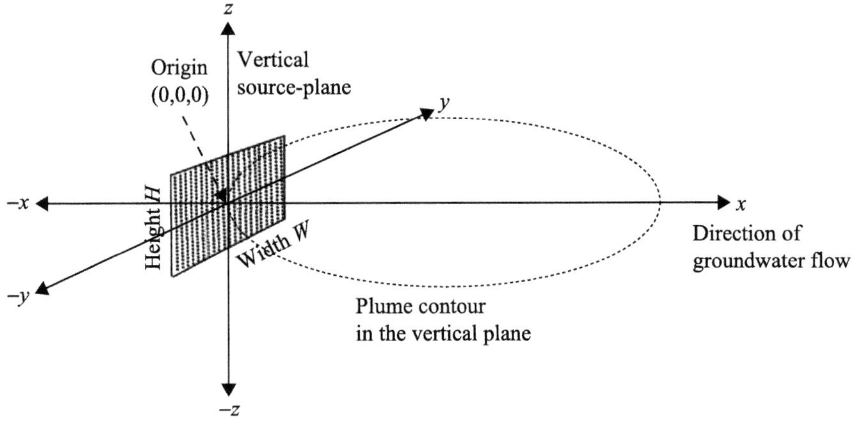

Fig. 5.2 Geometry and axis coordinate system for a vertical source-plane

(2) Material properties, including the hydraulic conductivity and dispersivity, are homogeneous.
(3) Partitioning between the dissolved and sorbed phases is instantaneous and reversible.
(4) The solute experiences first-order degradation at the same rate in both the dissolved and sorbed phases.

A range of boundary conditions can be employed to solve Eq. (5.1.52). For the scenario involving a constant source concentration, the "approximate" solution is a widely-used analytical solution for three-dimensional solute transport. The Green's function is used as the starting equation, which represents the solution for a point source in an infinite aquifer.

The governing equation for the Green's function is:

$$\begin{aligned} R\frac{\partial C}{\partial t} &= D_x\frac{\partial^2 C}{\partial x^2} + D_y\frac{\partial^2 C}{\partial y^2} + D_z\frac{\partial^2 C}{\partial z^2} - v\frac{\partial C}{\partial x} - R\lambda C \\ &\quad + \frac{M}{\theta}\delta(x-x')\,\delta(y-y')\delta(z-z')\delta(t-t') \end{aligned} \quad (5.1.54)$$

subject to boundary conditions:

$$C(\pm\infty, y, z, t) = 0 \quad (5.1.55)$$

$$C(x, \pm\infty, z, t) = C(x, y, \pm\infty, t) = 0 \quad (5.1.56)$$

and initial condition:

$$C(x, y, z, 0) = 0 \quad (5.1.57)$$

In Eq. (5.1.54), M is the mass released instantaneously at point (x', y', z') at time t', θ is aquifer porosity and δ is the Dirac delta function. The boundary conditions (Eqs. 5.1.55 and 5.1.56) incorporate the implicit assumption that the aquifer is infinitely wide and infinitely thick. The analytical solution is (Lenda and Zuber 1970):

$$\begin{aligned} C(x, y, z, t) &= \frac{M}{\theta}\frac{1}{8\pi^{\frac{3}{2}}}\frac{1}{\sqrt{\frac{D_x D_y D_z}{R}}}\frac{1}{(t-t')^{\frac{3}{2}}}\exp[-\lambda(t-t')] \\ &\quad \times \exp\left\{-\frac{\left[(x-x') - \frac{v(t-t')}{R}\right]^2}{\frac{4D_x(t-t')}{R}}\right\}\cdot\exp\left[-\frac{(y-y')^2}{\frac{4D_y(t-t')}{R}}\right] \\ &\quad \times \exp\left[-\frac{(z-z')^2}{\frac{4D_z(t-t')}{R}}\right] \end{aligned} \quad (5.1.58)$$

There are two ways to extend the solution for a point source to the case of a continuous plane-source of width W and height H. The first approach is by analytic integration and yields an expression that includes an integral that must be evaluated numerically. Some studies indicate the combination of two simpler solutions; one to account for dispersion in the x-direction and another for dispersion in the y and z directions. The transient solution proposed is:

$$\begin{aligned} C(x, y, z, t) = \frac{C_0}{8} \exp\left[\frac{vx}{2D_x}\left(1 - \sqrt{1 + \frac{4\lambda R D_x}{v^2}}\right)\right] \cdot \operatorname{erfc}\left(\frac{x - \frac{vt}{R}\sqrt{1 + \frac{4\lambda R D_x}{v^2}}}{2\sqrt{\frac{D_x t}{R}}}\right) \\ \times \left[\operatorname{erf}\left(\frac{z + \frac{H}{2}}{2\sqrt{\frac{D_z x}{v}}}\right) - \operatorname{erf}\left(\frac{z - \frac{H}{2}}{2\sqrt{\frac{D_z x}{v}}}\right)\right] \\ \times \left[\operatorname{erf}\left(\frac{y + \frac{W}{2}}{2\sqrt{\frac{D_y x}{v}}}\right) - \operatorname{erf}\left(\frac{y - \frac{W}{2}}{2\sqrt{\frac{D_y x}{v}}}\right)\right] \end{aligned} \quad (5.1.59)$$

where erf is the error function and $\operatorname{erfc}(x) = 1 - \operatorname{erf}(x)$.

The steady-state solution along the plume centreline is (Domenico 1987):

$$C(x, 0, 0, \infty) = C_0 \exp\left[\frac{vx}{2D_x}\left(1 - \sqrt{1 + \frac{4\lambda R D_x}{v^2}}\right)\right] \cdot \operatorname{erf}\left(\frac{W}{4\sqrt{\frac{D_y x}{v}}}\right) \cdot \operatorname{erf}\left(\frac{H}{4\sqrt{\frac{D_z x}{v}}}\right) \quad (5.1.60)$$

As the solution is symmetrical with respect to the surface $z = 0$ (Fig. 5.1), this surface can be interpreted either as the middle of the source located within an infinitely thick aquifer, or as the top of a semi-infinite aquifer. In this case, H is selected as twice the "true" height of the source.

As previously mentioned, the exact solution to the problem of a plane-source in an infinite domain can be achieved by integrating Eq. (5.1.59) over the release duration and dimensions of the source in the y and z directions. In order to represent a continuous release from a source with concentration C_0, the mass flux M must be defined. If the plane-source has a surface of constant concentration, the mass flux is composed of two components: the advective and dispersive components. However, due to the unknown value of the dispersive flux, it is neglected and only the advective mass flux is approximated:

$$\overline{M} = v\theta C_0 \quad (5.1.61)$$

The solution is:

$$C(x,y,z,t)=\frac{vC_0\exp\left(\frac{vx}{2D_x}\right)}{8\sqrt{\frac{\pi D_x}{R}}\int_0^t\frac{1}{\tau^{\frac{1}{2}}}\exp\left[-\left(\lambda+\frac{v^2}{4RD_x}\right)\tau-\frac{Rx^2}{4D_x\tau}\right]}\times\left[\operatorname{erfc}\left(\frac{y-\frac{W}{2}}{2\sqrt{\frac{D_y\tau}{R}}}\right)-\operatorname{erfc}\left(\frac{y+\frac{W}{2}}{2\sqrt{\frac{D_y\tau}{R}}}\right)\right]\times\left[\operatorname{erfc}\left(\frac{z-\frac{H}{2}}{2\sqrt{\frac{D_z\tau}{R}}}\right)-\operatorname{erfc}\left(\frac{z+\frac{H}{2}}{2\sqrt{\frac{D_z\tau}{R}}}\right)\right]d\tau \quad (5.1.62)$$

where τ is a dummy variable of integration.

An inadequacy of this solution arises due to the neglect of the dispersive component of the source mass flux, resulting in an underestimation of the concentration in the vicinity of the source. To precisely determine the concentration at the source with a prescribed value of C_0, it is essential to derive the solution for a semi-infinite domain in the x-direction. This is accomplished by applying the following source boundary condition to Eq. (5.1.52), subject to Eqs. (5.1.56) and (5.1.57):

$$\begin{array}{l}C(0,y,z,t)=C_0,\ \text{for}-\frac{W}{2}\le y\le\frac{W}{2}\text{and}-\frac{H}{2}\le z\le\frac{H}{2}\\ C(0,y,z,t)=0,\quad \text{for}|y|>\frac{W}{2}\text{and}|z|>\frac{H}{2}\end{array} \quad (5.1.63)$$

The solution, which is nearly identical to the previous one, is presented in Sagar (1982) and Wexler (1992):

$$C(x,y,z,t)=\frac{C_0x\exp\left(\frac{vx}{2D_x}\right)}{8\sqrt{\frac{\pi D_x}{R}}}\int_0^t\frac{1}{\tau^{\frac{3}{2}}}\exp\left[-\left(\lambda+\frac{v^2}{4RD_x}\right)\tau-\frac{Rx^2}{4D_x\tau}\right]\times\left[\operatorname{erfc}\left(\frac{y-\frac{W}{2}}{2\sqrt{\frac{D_y\tau}{R}}}\right)-\operatorname{erfc}\left(\frac{y+\frac{W}{2}}{2\sqrt{\frac{D_y\tau}{R}}}\right)\right]\times\left[\operatorname{erfc}\left(\frac{z-\frac{H}{2}}{2\sqrt{\frac{D_z\tau}{R}}}\right)-\operatorname{erfc}\left(\frac{z+\frac{H}{2}}{2\sqrt{\frac{D_z\tau}{R}}}\right)\right]d\tau \quad (5.1.64)$$

The corresponding steady-state solution is slightly awkward to compute as it involves two successive numerical integrations (one in y and another in z). It is more convenient to obtain the steady-state solution by computing Eq. (5.1.64) for large values of time. In the comparison provided below, we compare results obtained from Eqs. (5.1.59) and (5.1.64).

5.2 Continuous Time Random Walk (CTRW) Model

The continuous-time random walk (CTRW) transport equation is derived from the Fokker–Planck with memory equation (FPME), which characterizes the temporal change in probability density function of particle velocity as affected by both drag and random forces (Cortis and Berkowitz 2005; Jiménez-Hornero et al. 2005; Berkowitz et al. 2006; Bauget and Fourar 2008). In CTRW, the Laplace transformed concentration function $\tilde{C}(x, w)$, is given by the one-dimensional form of the FPME

$$w\tilde{C}(x,w) - C_0(x) = -\tilde{M}(w)\left[u_\psi \frac{\partial}{\partial x}\tilde{C}(x,w) - D_{\mathrm{CTRW}}\frac{\partial^2}{\partial x^2}\tilde{C}(x,w)\right] \quad (5.2.1)$$

where w is the Laplace variable, u_ψ and D_{CTRW} are transport velocity and dispersion in the framework of the CTRW, respectively. In Eq. (5.2.1), the memory function $\tilde{M}(w)$ captures the anomalous or non-Fickian transport induced by local heterogeneity or process and the corresponding formulation is given by

$$\tilde{M}(w) = \bar{t}w\frac{\tilde{\psi}(w)}{1-\tilde{\psi}(w)} \quad (5.2.2)$$

where $\bar{t}$ is some characteristic time and $\tilde{\psi}(w)$ is the transition rate probability which is the core of the CTRW model. In general, there are three different models for $\tilde{\psi}(w)$: The asymptotic model, the truncated power law (TPL) model, and the modified exponential model. A detail for three different models of $\tilde{\psi}(w)$ can be found in the reference of Cortis and Berkowitz (2004). Here, we only used the TPL model in the framework of the CTRW,

$$\tilde{\psi}(w) = (1 + t_2wt_1)^{\beta}\exp(t_1w)\frac{\Gamma(-\beta, t_2^{-1} + t_1w)}{\Gamma(-\beta, t_2^{-1})} \quad (5.2.3)$$

where t_1 represents the lower limit time when the power law behavior begins, t_2 is the cut-off time describing the Fickian behavior dominates, and Γ represents the incomplete Gamma function.

The parameter β indicates the different types of anomalous transport. As Cortis et al. (2004) pointed out, $0 < \beta < 1$ indicated a highly anomalous transport; $1 < \beta < 2$ was associated with a moderately anomalous transport; $\beta > 2$ corresponded to the Fickian transport.

5.3 Mobile-Immobile (MIM) Model

The single-rate mobile-immobile (MIM) model differs from the Fickian model in that it assumes Fickian transport in the mobile domain and diffusion-driven mass exchange between the mobile and immobile domains. In the case of conservative solute transport, the MIM model is represented by the following equation:

$$\theta_{\mathrm{m}}\frac{\partial C_{\mathrm{m}}}{\partial \mathrm{t}} = \theta_{\mathrm{m}} D_{\mathrm{f,m}}\frac{\partial^2 C_{\mathrm{m}}}{\partial x^2} - \overline{u}\theta_{\mathrm{m}}\frac{\partial C_{\mathrm{m}}}{\partial x} - \alpha(C_{\mathrm{m}} - C_{\mathrm{im}}) \tag{5.3.1}$$

$$\theta_{\mathrm{im}}\frac{\partial C_{\mathrm{im}}}{\partial t} = \alpha(C_{\mathrm{m}} - C_{\mathrm{im}}) \tag{5.3.2}$$

where C_{m} and C_{im} represent the solute concentrations in the mobile and immobile domains, respectively; θ_{m} and θ_{im} represent the water contents in the mobile and immobile domains, respectively, and the sum of θ_{m} and θ_{im} is equal to one for a saturated 2D fracture; $D_{\mathrm{f,m}}$ is the dispersion coefficient and α is the mass transfer rate between the mobile and immobile domains (T^{-1}).

A dimensionless $\beta = \theta_{\mathrm{m}}/(\theta_{\mathrm{im}}+\theta_{\mathrm{m}})$ is introduced as the mobile water fraction. It is important to note that our decision to apply the MIM model is founded on the belief that the potential eddy-controlled domain can be viewed as an immobile domain, rather than considering the fracture matrix as the immobile domain. Consequently, we assume that the fracture matrix is impermeable.

5.4 Spatial Moment

Spatial moments are a useful tool to describe the distribution of solute concentration. The zeroth moment corresponds to the total mass of solute in the domain:

$$M_0(t) = \int_V C(x,t)\mathrm{d}V \tag{5.4.1}$$

The first spatial moment $M_{1,1}(t)$ represents the location of the plume centroid in the main flow direction, i.e., the longitudinal direction:

$$M_{l,1}(t) = \frac{\int_V C(x,t)x\mathrm{d}V}{M_0(t)} \tag{5.4.2}$$

The second central spatial moment $M_{2,1,1}(t)$ represents the mean square displacement from the plume centroid in the longitudinal direction:

$$M_{2,l,l}(t) = \int_V C(x,t)[x - M_{1,1}(t)][x - M_{1,1}(t)]\mathrm{d}V \tag{5.4.3}$$

In fact, $M_{2,1,1}(t)$ is a measure of spreading and dispersion of the plume.

The spatial moment method can serve as a tool to verify the analytical solution for Taylor dispersion. Taylor's classic study showed that for solute transport through parallel plates, where the fluid experiences stratified flow, the effective solute transport eventually becomes one-dimensional, and the dispersion induced by shear flow can be quantified by an effective longitudinal dispersion coefficient (i.e., Taylor dispersion coefficient) and mean flow velocity $\overline{u}$. The Taylor dispersion coefficient for an ideal 2D parallel fracture can be expressed as,

$$D_{\text{Taylor}} = \frac{\overline{u}^2\overline{b}^2}{210D_{\text{m}}} \tag{5.4.4}$$

The length and height of the parallel plate were $L = 200$ mm and $H = 0.5$ mm, the ratio of length to height is $L/H = 400$, and the corresponding asymptotic time is $\tau = b^2/4D_{\text{m}} = 62.5$ s. The average flow velocity in the fracture is $\overline{u} = 6.17 \times 10^{-5}\text{m/s}$.

Figure 5.3 demonstrates that the dispersion coefficient prior to the progressive time is not a constant, but instead, is a function dependent on time. Based on the physical interpretation of the spatial moment, the dispersion coefficient in the x-direction can be expressed as the rate of change of the second-order moment in the x-direction,

$$2D_x = \lim_{t\to\infty} \frac{\mathrm{d}M_{2,xx}}{\mathrm{d}t} \tag{5.4.5}$$

The total mass of solute in the calculation area is calculated by using the spatial moment of solute cluster

$$M = \int_\Omega C(x,t)\mathrm{d}\Omega \tag{5.4.6}$$

First moment in x-direction:

$$M_{1,x}(t) = \int_\Omega C(x,t)x\mathrm{d}\Omega \tag{5.4.7}$$

Second moment in x-direction:

$$M_{2,xx}(t) = \frac{\int_\Omega C(x,t)(x - M_{1,x}/M)(x - M_{1,x}/M)\mathrm{d}\Omega}{M} \tag{5.4.8}$$

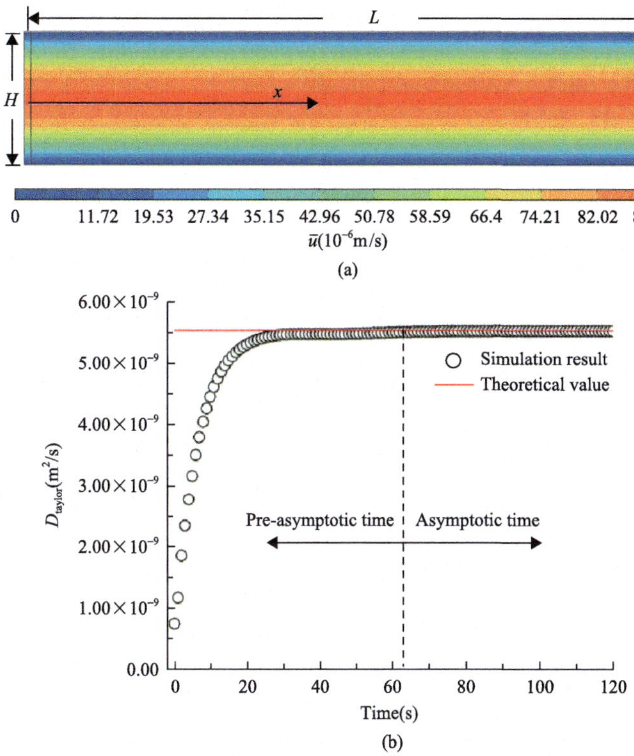

Color picture

Fig. 5.3 The schematic diagram of flow and Taylor dispersion coefficient **a** the shear flow in a parallel plate; **b** the dynamic variation process of Taylor dispersion coefficient

Figure 5.3 depicts that the dispersion coefficient is not constant before the progressive time, and it varies with time. The spatial moment method can be used to express the dispersion coefficient in the x-direction by the second-order moment's rate of change in the same direction. The calculated dispersion coefficient at the asymptotic time using the second-order moment in the x-direction is consistent with the theoretical value of Taylor dispersion coefficient.

5.5 Scalar Dissipation Rate (SDR) and Dilution Index

5.5.1 Scalar Dissipation Rate (SDR)

The scalar dissipation rate (SDR) is an integral measure of concentration gradients and can be used to characterize mixing. In recent years, studies have focused on the evolution and scaling properties of SDR during solute transport in porous media.

However, investigations into SDR evolution in rough fractures have been limited. This limitation has prompted our study to fill this gap in knowledge. The SDR of a conservative scalar can be expressed as follows:

$$\chi(t) = \int\limits_{\Omega} D_{\mathrm{m}} \nabla C(x,t) \cdot \nabla C(x,t)\, \mathrm{d}x \tag{5.5.1}$$

In order to calculate the SDR using Eq. (5.5.1), it is necessary to determine the local concentration gradients. However, due to the occurrence of sharp concentration gradients over small distances in a relatively heterogeneous flow field, obtaining an accurate quantification of concentration gradients requires a very fine numerical discretization for both the flow and concentration fields, resulting in a significant computational cost. To mitigate this issue, Le Borgne et al. (2010) demonstrated that the SDR can be approximated using the concentration second moment, which is the integral of the squared concentrations. After multiplying Eq. (5.5.1) by $C(x,t)$ and integrating over the entire domain, we have

$$\begin{aligned} &\frac{1}{2}\frac{\partial}{\partial t}\int\limits_{\Omega} C(x,t)^2 d\Omega + \frac{1}{2}\int\limits_{\Omega} \nabla \cdot \left[uC(x,t)^2\right] \mathrm{d}\Omega \\ &= \frac{1}{2} D_{\mathrm{m}} \int\limits_{\Omega} \nabla \cdot \nabla C(x,t)^2 \mathrm{d}\Omega - \int\limits_{\Omega} D_{\mathrm{m}} \nabla C(x,t) \cdot \nabla C(x,t) \mathrm{d}\Omega \end{aligned} \tag{5.5.2}$$

Given that the fractured domain is infinite, the mass flux out of the domain is assumed to be zero. Additionally, the flow field is assumed to be divergence-free. In this case, the terms in Eq. (5.5.2) that involve a divergence operator become zero. Then we obtain

$$\chi(t) = -\frac{1}{2}\frac{\mathrm{d}M^2(t)}{\mathrm{d}t} = D_{\mathrm{m}} \int\limits_{\Omega} \nabla \cdot \nabla C(x,t)^2 \mathrm{d}\Omega \tag{5.5.3}$$

where $M^2(t)$ is the concentration second moment and can be expressed as

$$M^2(t) = \int\limits_{\Omega} C\,(x,t)^2 \mathrm{d}\Omega \tag{5.5.4}$$

Le Borgne et al. (2010) reported that the results from using Eq. (5.5.3) are more accurate than from using Eq. (5.5.1) directly and the calculation for Eq. (5.5.3) is computationally more efficient. So the temporal mixing state is obtained from Eq. (5.5.3).

In a 1D homogeneous domain with zero velocity, the solute concentration distribution can be described using Fick's Law of diffusion. In the absence of any reaction, the corresponding analytical solution for the concentration distribution can be obtained by,

$$C_0(x,t) = \frac{m_0}{\sqrt{4\pi D_m t}} \exp\left(-\frac{x^2}{4\pi D_m}\right) \tag{5.5.5}$$

The power-law function of the dimensionless time was fitted to Fig. 5.4 over the entire temporal range. By integrating the square of Eq. (5.5.5) over all domains, the concentration second moment can be expressed as

$$\int_{\Omega} C_0(x,t)^2 d\Omega = \frac{m_0^2}{\sqrt{8\pi D_m t}} = M_0(t) \tag{5.5.6}$$

From Eq. (5.5.3), the analytical 1D SDR solution can thus be expressed as

$$\chi_0(t) = -\frac{1}{2}\frac{dM_0(t)}{dt} = \frac{1}{8}\frac{m_0^2}{\sqrt{2\pi D_m}} t^{-\frac{3}{2}} \tag{5.5.7}$$

5.5.2 Dilution Index

The concept of dilution index was first introduced by Kitanidis (1994). The dilution index represents the exponential of the Shannon entropy of a concentration probability distribution and is given as

$$E(t) = \exp\left\{-\int_{-\infty}^{+\infty}\int_{-\infty}^{+\infty} p_g(x,y,t)\ln\left[p_g(x,y,t)\right]dxdy\right\} \tag{5.5.8}$$

where $p_g(x, y, t)$ is the concentration probability function of the solute and defined by

$$p_g(x,y,t) = \frac{C(x,y,t)}{\int_{-\infty}^{+\infty}\int_{-\infty}^{+\infty} C(x,y,t)dxdy} \tag{5.5.9}$$

For a 2D fracture, the maximum value of the dilution index, $E(t)_{max}$, is technically equal to the area of the entire fracture, representing a uniform concentration distribution in the fracture. The dilution process from Eq. (5.5.8) is considered as "the act of distributing solute mass over a larger volume".

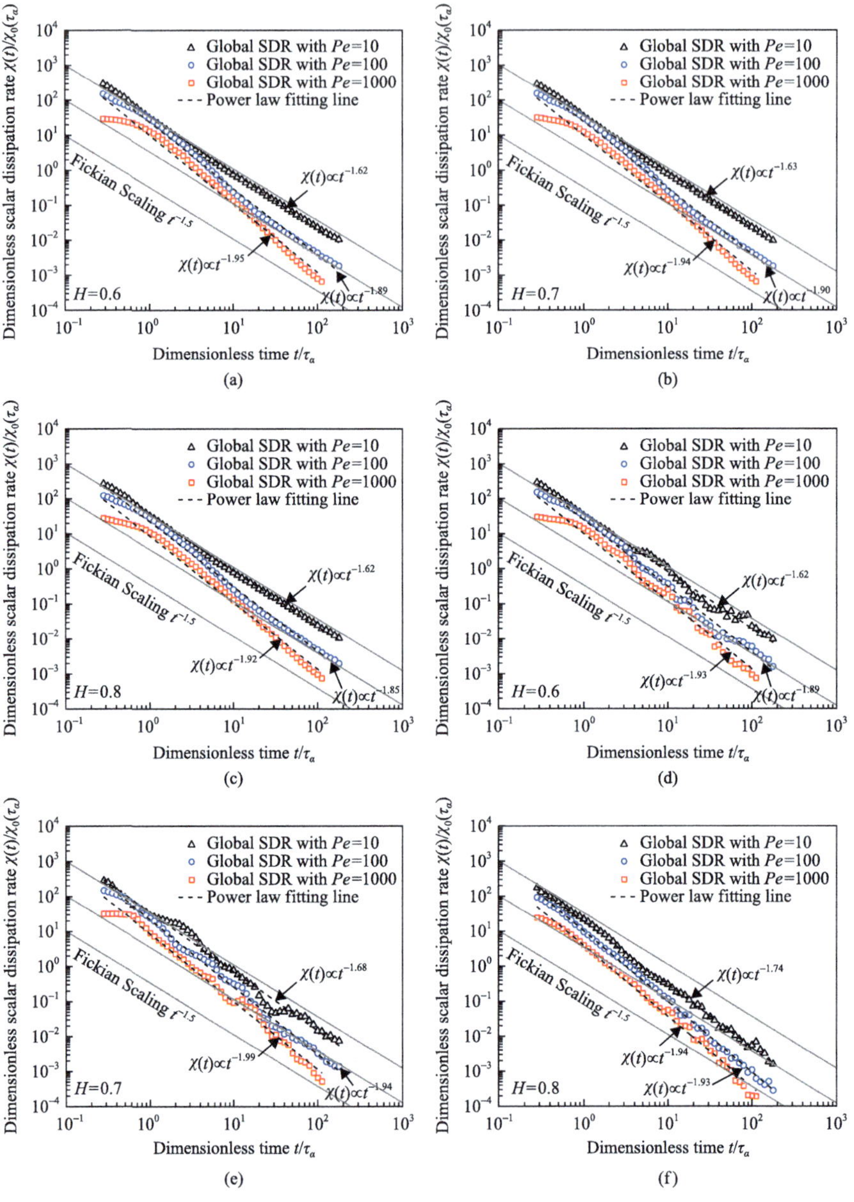

Fig. 5.4 Scalar dissipation rate estimated in the constant-aperture fracture (**a**), (**b**), (**c**) and variable-aperture fracture (**d**), (**e**), (**f**) with $H = 0.6$, 0.7, and 0.8 for $Pe = 10$, 100, and 1000, respectively

It is worth noting that the dilution index described in Eq. (5.5.8) can be interpreted as the effective volume occupied by a solute plume. This characteristic is valuable for analyzing the solute dilution process when the solute is injected as a single slug, as well as for quantifying the uniformity of concentration distribution within the fracture when the solute is continuously injected, due to the property of the Shannon entropy in Eq. (5.5.8).

References

Bauget F, Fourar M (2008) Non-Fickian dispersion in a single fracture. J Contam Hydrol 100(3–4):137–148

Berkowitz B, Cortis A, Dentz M et al (2006) Modeling non-Fickian transport in geological formations as a continuous time random walk. Rev Geophys 44(2). https://doi.org/10.1029/2005rg000178

Cortis A, Berkowitz B (2004) Anomalous transport in "classical" soil and sand columns. Soil Sci Soc Am J 68(5):1539–1548

Cortis A, Berkowitz B (2005) Computing "anomalous" contaminant transport in porous media: the CTRW MATLAB toolbox. Ground Water 43(6):947–950

Cortis A, Gallo C, Scher H et al (2004) Numerical simulation of non-Fickian transport in geological formations with multiple-scale heterogeneities. Water Resour Res 40(4):W04209

Domenico PA (1987) An analytical model for multidimensional transport of a decaying contaminant species. J Hydrol 91(1–2):49–58

Jiménez-Hornero F, Giráldez J, Laguna A et al (2005) Continuous time random walks for analyzing the transport of a passive tracer in a single fissure. Water Resour Res 41(4). https://doi.org/10.1029/2004WR003852

Kitanidis PK (1994) The concept of the dilution index. Water Resour Res 30(7):2011–2026

Le Borgne T, Dentz M, Bolster D et al (2010) Non-Fickian mixing: temporal evolution of the scalar dissipation rate in heterogeneous porous media. Adv Water Resour 33(12):1468–1475

Lenda A, Zuber A (1970) Tracer dispersion in groundwater experiments. Isotope Hydrology, Vienna, pp 619–641

Sagar B (1982) Dispersion in three dimensions: approximate analytic solutions. J Hydraul Div 108(1):47–62

Wexler EJ (1992) Analytical solutions for one-, two-, and three-dimensional solute transport in ground-water systems with uniform flow. US Government Printing Office, Washington D.C.

Chapter 6
Numerical Methods of Mass Transfer Process in Fractured Media

In the previous chapter we saw that physical phenomena of interest to us can be described by a general scalar transport equation. In this chapter, we examine numerical methods for solving this type of equation and introduce the different numerical methods for mass transfer. We also explore how to characterize our numerical methods in terms of accuracy, consistency, stability, and convergence.

6.1 Lattice Boltzmann Method

The lattice Boltzmann method (LBM) has gained widespread acclaim recent years. The LBM is a powerful technique for the computational modeling of many different complicated fluid flow issues, such as single and multiphase flow in complex geometries. This method naturally accommodates a variety of boundary conditions such as the pressure drop across the interface between two fluids and wetting effects at a fluid–solid interface and a constant velocity or pressure boundary. It is an approach that bridges microscopic phenomena with the continuum macroscopic equations. On the other hand, the application of LBM in simulating fluid–solid interactions, modeling the behavior of fluid/fluid interactions, and managing irregular flow paths is acknowledged due to the algorithm's accuracy and simplicity. Several models are provided in LBM for the analysis of immiscible two-phase flow. They are the chromodynamics model (Gunstensen and Rothman 1991), the pseudo-potential model (Shan and Chen 1993, 1994; Shan and Doolen 1994; Martys and Chen 1996), free energy model (Swift et al. 1995, 1996) and He-Shan-Doolen model (He et al. 1999).

The MCSC LBM can be implemented to study 2D steady-state immiscible two-phase flow system. For this, two distribution functions are employed to represent WP (wetting phase) and NWP (non-wetting phase), respectively. The evolution equation of each distribution function satisfies the following lattice Boltzmann equation

Z. Dou et al., *Mass Transfer Dynamics of Contaminants in Fractured Media*,
https://doi.org/10.1007/978-981-99-9187-7_6

$$f_i^{\sigma}(x + e_i \Delta t, t + \Delta t) = f_i^{\sigma}(x, t) - \frac{\Delta t}{\tau^{\sigma}}\left[f_i^{\sigma}(x, t) - f_i^{\sigma(\text{eq})}(x, t)\right] \quad (6.1.1)$$

where $f_i^{\sigma}(x, t)$ is the fluid particle distribution function, e_i is the direction of particle motivation, and σ denotes either the WP or NWP, τ^{σ} is the relaxation time which is related to the kinematic viscosity as $\nu^{\sigma} = c_s^2(\tau^{\sigma} - 0.5\Delta t)$. The relaxation time τ is the most important model parameter in LBM. It characterizes the time-scale behavior of fluid particle collisions and determines the lattice fluid viscosity in the lattice scale. $f_i^{\sigma(\text{eq})}(x, t)$ is the local equilibrium distribution and defined as

$$f_i^{\sigma(\text{eq})}(x, t) = \omega_i \rho^{\sigma}\left[1 + \frac{e_i \cdot u^{\sigma(\text{eq})}}{c_s^2} + \frac{(e_i \cdot u^{\sigma(\text{eq})})^2}{2c_s^4} - \frac{(u^{\sigma(\text{eq})})^2}{2c_s^2}\right] \quad (6.1.2)$$

where c_s is the lattice sound speed, and $c_s = \frac{c}{\sqrt{3}}$, where $c = \frac{\Delta x}{\Delta t}$ is the ratio of lattice spacing Δx and time step Δt. In our simulations, one lattice unit (Δx) is defined as 1 Lu and one time step (Δt) is defined as 1 lattice unit. In this article, all the variables are in lattice units. In Eq. (6.1.2), the weights ω_i, for the D2Q9 model (Fig. 6.1) (Qian et al. 1992), are given by

$$\omega_i = \begin{cases} \frac{4}{9}, & i = 0 \\ \frac{1}{9}, & i = 1, 2, 3, 4 \\ \frac{1}{36}, & i = 5, 6, 7, 8 \end{cases} \quad (6.1.3)$$

In Eqs. (6.1.1) and (6.1.2), the e_i's are the discrete velocities. For the D2Q9 model, they are given by

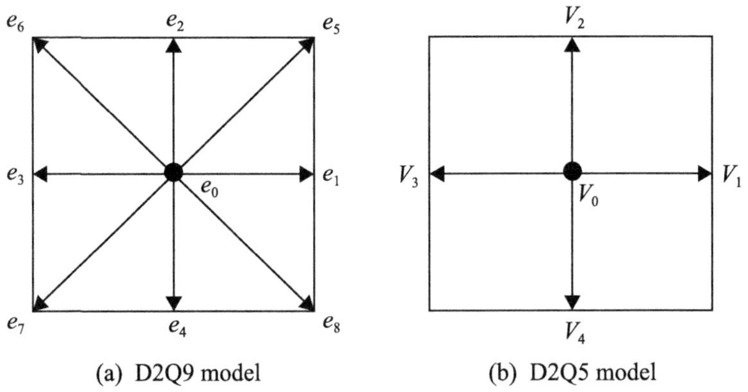

Fig. 6.1 Lattice geometries and velocity vectors of a D2Q9 model and a D2Q5 model

$$[e_0, e_1, e_2, e_3, e_4, e_5, e_6, e_7, e_8] = c\begin{bmatrix} 0 & 1 & 0 & -1 & 0 & 1 & -1 & -1 & 1 \\ 0 & 0 & 1 & 0 & -1 & 1 & 1 & -1 & -1 \end{bmatrix} \tag{6.1.4}$$

The local mass densities ρ^σ for each phase are obtained from

$$\rho^\sigma = \sum_i f_i^\sigma \tag{6.1.5}$$

The macroscopic velocity $u^{\sigma(\mathrm{eq})}$ in Eq. (6.1.2) is given by

$$u^{\sigma(\mathrm{eq})} = u + \Delta u = u + \frac{\tau^\sigma F^\sigma}{\rho^\sigma} \tag{6.1.6}$$

$$u = \frac{\sum_\sigma \frac{1}{\tau^\sigma} \sum_i f_i^\sigma e_i}{\sum_\sigma \frac{1}{\tau^\sigma} \rho^\sigma} \tag{6.1.7}$$

In Eqs. (6.1.6) and Eq. (6.1.7), u is the common averaged velocity of all the fluid components in the absence of any additional forces. Δu is a change in velocity. In the MCSC LBM model, the effect of total force is incorporated through adding acceleration into the velocity field. In our simulations, the total force includes the fluid–fluid cohesion F_{c}^σ, the fluid–solid adhesion F_{ads}^σ and the external body force G. In Eq. (6.1.6), $F^\sigma = F_{\mathrm{c}}^\sigma + F_{\mathrm{ads}}^\sigma + G$, the external body force is a steady body force and is corresponding force to the pressure gradient. It should be noted that the initial values of particle distribution function of each component σ are usually assumed as calculated local equilibrium density distribution for the same component based on initial macroscopic fluid density and velocity.

In the MCSC model, only the coupling interaction between the closest neighbors and the next-nearest neighbors is taken into account when defining the fluid–fluid interaction in Eq. (6.1.8). (Shan and Doolen 1994; Martys and Chen 1996; Kang et al. 2002; Huang and Lu 2009),

$$F_{\mathrm{c}}^\sigma(x,t) = -G_{\mathrm{c}}\rho^\sigma(x,t)\sum_i \omega_i \rho^{\overline{\sigma}}(x + e_i\Delta t, t)e_i \tag{6.1.8}$$

where the σ and $\overline{\sigma}$ denote the WP and the NWP, respectively, and G_{c} is a parameter that controls the strength of fluid–fluid interaction. The surface force between a fluid component and a solid phase, $F_{\mathrm{ads}}^\sigma(x,t)$, is defined as (Shan and Chen 1993, 1994; Shan and Doolen 1994; Benzi et al. 2006; Huang and Lu 2009)

$$F_{\mathrm{ads}}^\sigma(x,t) = -G_{\mathrm{ads}}^\sigma \rho^\sigma(x,t)\sum_i \omega_i s(x + e_i\Delta t, t)e_i \tag{6.1.9}$$

where $s(x + e_i\Delta t, t)$ is an indicator function that is equal to 1 or 0 for a solid or a fluid domain node, respectively, and G_{ads}^σ is a parameter that controls the strength between each fluid and a wall. The macroscopic physical quantities of the flow, such as density in Eq. (6.1.5), and the actual whole fluid velocity V is defined as

$$V = u + \frac{\sum_{\sigma} F^{\sigma}}{2\sum_{\sigma} \rho^{\sigma}} \tag{6.1.10}$$

The pressure $P(x)$ of the whole fluid is given as

$$P(x) = c_s^2 \sum_{\sigma} \rho(x) + \frac{G_c}{3}\rho^{\sigma}(x)\rho^{\overline{\sigma}}(x) \tag{6.1.11}$$

LBM has a remarkable ability to simulate single and multiphase fluids. A rich variety of behaviors, including unsteady flows, phase separation, evaporation, condensation, cavitation, solute and heat transport, buoyancy, and interactions with surfaces can readily be simulated. Persistent metastable states can be realized. Another exciting use of LBM is for the simulation of shallow flows with the shallow water equations.

In Fig. 6.2, the flow direction is from left to right, the light grey domain represents WP, the black domain represents NWP and the dark grey domain represents solid, which is similar with oncoming figures. Figure 6.2a, b show the evolutions of the phase distribution patterns in the heterogeneous and homogeneous porous media, respectively. The WP covers most of the solid surface, in contrast, the NWP flows in form of small droplets and is not continuous. When $S_w = 0.3$, the NWP is relatively continuous and can travel in the connected channel, while the WP is discontinuous and covers the solid surface. For the cases of $S_w = 0.6$ and $S_w = 0.9$, the NWP forms several large droplets and is discontinuous. These droplets are trapped in pores due to the capillary resistance and immobile property. Especially, the distributions of NWP in the heterogeneous porous media, in general, are more continuous than that in the homogeneous porous media, while the distributions of the WP in heterogeneous porous media are more dispersive than that in homogeneous porous media. These variations lead to the fact that the RP of NWP with heterogeneous porous media is higher than that with homogeneous porous media.

6.2 Immiscible Two-Phase Transport Model: Phase Field Method

An immiscible two-phase transport (ITPT) model may be employed by connecting modified Navier–Stokes equation (including a field-dependent surface tension force) with advective Cahn–Hilliard equation to simulate immiscible two-phase flow transport and to obtain the residual non-wetting phase distribution. The modified Navier–Stokes equation (Eq. 6.2.1), which represents the conservation of momentum for the fluid flow, and the continuity equation (Eq. 6.2.2) directly solve the fluid flow in porous media for an isothermal, incompressible, and Newtonian fluid.

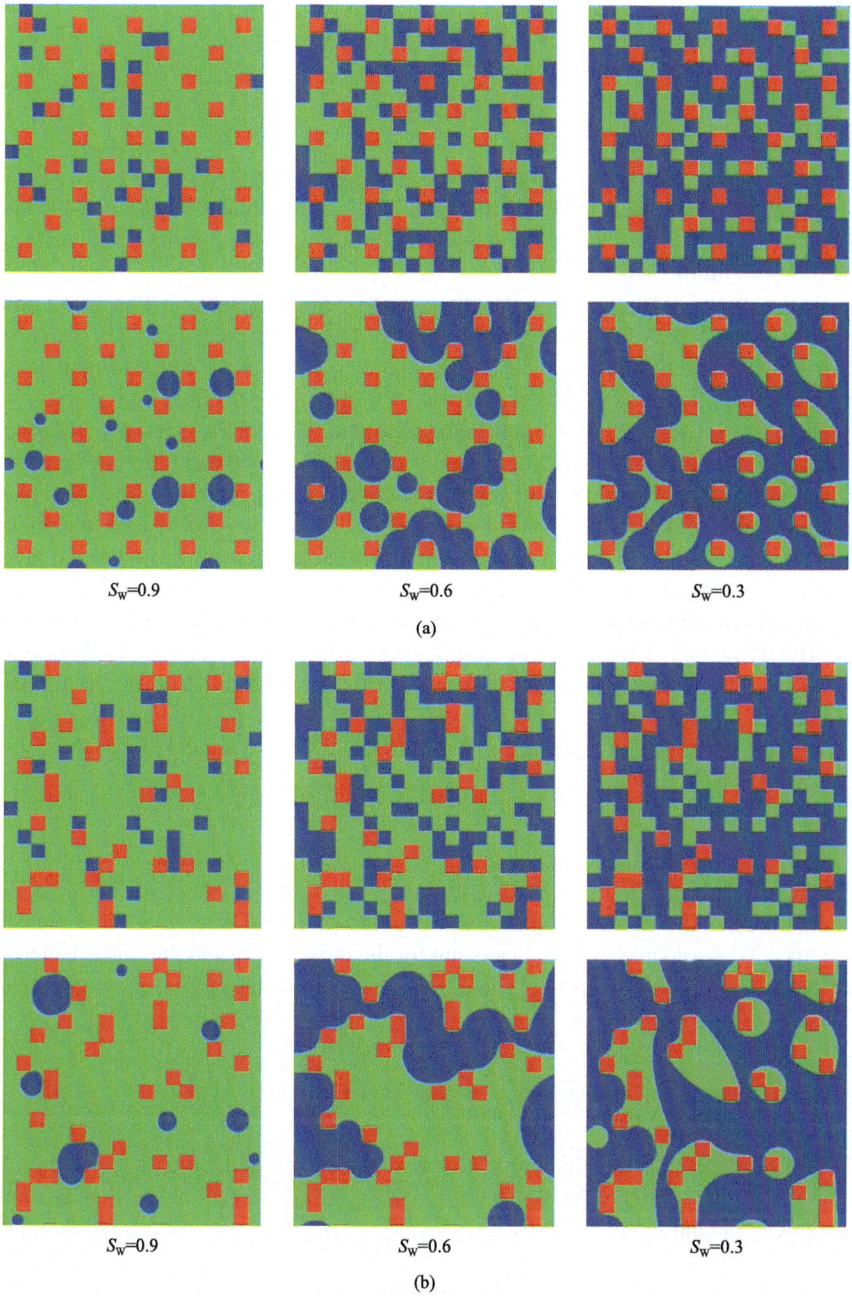

Fig. 6.2 LBM results for the spatial distribution of the WP and NWP in heterogeneous and homogeneous porous media

$$\rho\frac{\partial u}{\partial t}+\rho u\cdot\nabla u=-\nabla p+\nabla\cdot\left[\mu(\nabla u+\nabla u^{\mathrm{T}})\right]+F_{\mathrm{st}} \tag{6.2.1}$$

$$\nabla\cdot u=0 \tag{6.2.2}$$

where ρ is the density of fluid (ML^{-3}), p is the fluid pressure ($ML^{-1}T^{-2}$), u is the flow velocity of fluid (LT^{-1}), t is the time (T), μ is the dynamic viscosity of fluid ($ML^{-1}T^{-1}$), and F_{st} is the surface tension force acting at the interface between two immiscible fluids ($ML^{-2}T^{-2}$).

In the absence of the description of fluid/fluid interface movement, Eqs. (6.2.1) and (6.2.2) is incapable of simulating the immiscible two-phase transport. Alternatively, to capture the fluid/fluid interface during the transport, an interface tracking method, phase field method (PFM), was used to track the interface movement and couple with Eqs. (6.2.1) and (6.2.2). The PFM is thermodynamic-based and assumes the interface between two immiscible fluids as a physically diffuse layer with thin but non-zero thickness. The phase-field parameter, ϕ, associating with a continuous variation over the thin interfacial layer but uniformity in both bulk fluids, was introduced to separate the two immiscible fluids and track the interface. For instance, the interfacial layer can be considered as the region where the ϕ varies from -1 to 1, while the two immiscible fluids can be presented by $\phi=1$ and $\phi=-1$, respectively.

The dynamics of the ϕ can be described by a free-energy density model for an isothermal immiscible two-phase flow system (Villanueva and Amberg 2006). The free energy is given as

$$F(\phi)=\int\limits_{\Omega}\left(f_D(\phi)+\frac{1}{2}\lambda|\nabla\phi|^2\right)\mathrm{d}\Omega \tag{6.2.3}$$

where Ω represents the space occupied by the two immiscible fluids and λ is the magnitude of the mixing energy (MLT^{-2}). The terms on the right-hand side of Eq. (6.2.3), $f_D(\phi)$ and $\frac{1}{2}\lambda|\nabla\phi|^2$, account for the bulk energy and the surface energy, respectively. $f_D(\phi)$ is defined by considering a double-well potential (Badalassi et al. 2003),

$$f_D(\phi)=\frac{\lambda}{4\varepsilon^2}(\phi+1)^2(\phi-1)^2 \tag{6.2.4}$$

where ε represents the thickness of the interface between the two immiscible fluids (L). The surface tension coefficient, σ (MT^{-2}), can be related to both parameters λ and ε

$$\sigma=\frac{2\sqrt{2}}{3}\frac{\lambda}{\varepsilon} \tag{6.2.5}$$

The chemical potential, $\xi(\phi)$, is introduced as (Yue et al. 2004)

$$\xi(\phi) = \frac{\delta F}{\delta \phi} = \lambda\left[-\nabla^2\phi + \frac{\phi(\phi^2 - 1)}{\varepsilon^2}\right] \tag{6.2.6}$$

where $\frac{\delta}{\delta\phi}$ denotes the variational differentiation of the free energy (Eq. 6.2.3) of the interface region with respect to ϕ. By considering the mass flux as proportional to the gradient of $\xi(\phi)$, the advective Cahn–Hilliard equation is given as

$$\frac{\partial \phi}{\partial t} + u \cdot \nabla\phi = \nabla \cdot \left(\frac{\gamma\lambda}{\varepsilon^2}\nabla\psi\right) \tag{6.2.7}$$

$$\psi = -\nabla \cdot \varepsilon^2\nabla\phi + (\phi^2 - 1)\phi \tag{6.2.8}$$

where γ is the mobility parameter ($M^{-1}L^3T$) and denotes the moving velocity of the interface under a unit driving force. γ can be considered as a function of ε

$$\gamma = \chi\varepsilon^2 \tag{6.2.9}$$

where χ is the mobility tuning parameter. Since χ defines the interfacial relaxation time and determines the time scale of the Cahn–Hilliard diffusion, the value of χ is crucial for the numerical stability of the immiscible two-phase transport model. The value of χ must be high enough to keep the interface thickness constant and low enough not to damp the advection (Jacqmin 1999).

Recalling the surface tension force F_{st} in the modified Navier–Stokes equation (Eq. 6.2.1), F_{st} now can be related to the phase-field parameter ϕ and given as

$$F_{\mathrm{st}} = \xi(\phi)\nabla\phi \tag{6.2.10}$$

Both density and viscosity in Eq. (6.2.1) are the functions of the phase-field parameter and defined as

$$\rho = \frac{1+\phi}{2}\rho_{\mathrm{nw}} + \frac{1-\phi}{2}\rho_{\mathrm{w}} \tag{6.2.11}$$

$$\mu = \frac{1+\phi}{2}\mu_{\mathrm{nw}} + \frac{1-\phi}{2}\mu_{\mathrm{w}} \tag{6.2.12}$$

The immiscible two-phase transport model is validated by simulating the process of two-dimensional bubbles rising in liquid columns due to buoyancy force and comparing the results to those of Hysing et al. (2009). Figure 6.3 illustrates the calculated bubble mass center and rising bubble velocity at various periods. The simulation results are in strong agreement with those of Hysing et al. demonstrating the accuracy of numerical models.

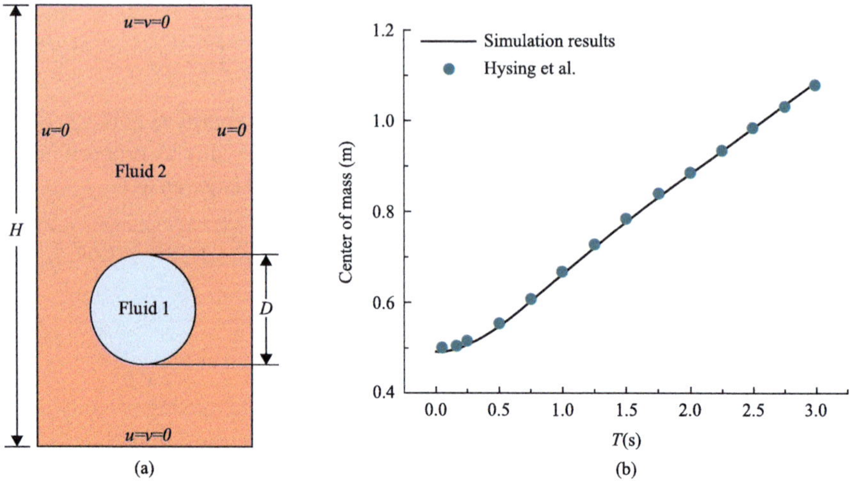

Fig. 6.3 Validation of the immiscible two-phase transport model **a** initial configuration and boundary conditions of the bubble rising case; **b** time evolution of the mass center

6.3 Pore-Scale Aqueous Solute Transport Model

The aqueous- miscible tracer transport in residual unsaturated porous media is simulated using the pore-scale aqueous solute transport model. The typical groundwater flow velocity ranges from 0.1 m/d to 10 m/d and is laminar. Under isothermal, incompressible, Newtonian, and steady-state flow conditions, the aqueous flow field is immediately solved using the continuity equation (Eq. 6.2.2) and the N-S equation (Eq. 6.3.1)

$$\rho u \cdot \nabla u = -\nabla p + \nabla \cdot \left[\mu(\nabla u + \nabla u^{\mathrm{T}})\right] \tag{6.3.1}$$

For the concentration field of the aqueous-miscible tracer, the advection–diffusion equation is used

$$\frac{\partial C}{\partial t} = -\nabla(uC) + D_{\mathrm{m}}\nabla^2 C, \quad x \in \Omega_f \tag{6.3.2}$$

where C is the aqueous-miscible tracer concentration (ML^{-3}), D_{m} is the molecular diffusion coefficient ($\mathrm{L}^2\mathrm{T}^{-1}$), and the 2D velocity tensor $u(\mathrm{LT}^{-1})$is determined by the numerical solution of Eqs. (6.2.2) and (6.3.1).

6.4 Coupling Strategy

The void space in porous media is initially saturated by the non-wetting phase liquid. Trichloroethylene (TCE) is assumed to be the non-wetting phase liquid and water is introduced as the wetting phase liquid. Water is injected into porous media at three injection rates to displace the TCE. In immiscible fluid–fluid displacement process, Lenormand et al. (1988) proposes the pioneered phase diagram to distinguish displacement patterns by two dimensionless numbers [e.g., capillary number $Ca(Ca = \mu_i v_i/\sigma)$ and viscosity ratio $M(M = \mu_i/\mu_d)$, where μ_i and μ_d are the viscosity of the invading and defending fluids, respectively, v_i is the characteristic velocity of the invading fluid, and σ is the interfacial tension]. Ca quantifies the relative magnitude of the viscous force to the capillary force. To obtain the residual TCE distribution patterns, the displacement processes are considered. The IPTP model is first implemented with three different capillary numbers (Ca) ranging from lgCa = –4.12 to –6.60, while the viscosity ratio (M) is constant and the lg$M$ is equal to 0.64. The left and right boundaries of the porous medium are the inlet and outlet boundaries, respectively. Water is injected along the inlet boundary at three specific flow rates (e.g., $u = 0.60$ m/s, 0.15 m/s, and 0.05 m/s). The outlet boundary is set as zero pressure boundary condition. The non-slip boundary condition is applied to the immiscible interfaces between TCE, water and soil.

Since the residual TCE was trapped, the aqueous space for the tracer transport varies with the residual saturation of TCE. Once the residual TCE distribution is obtained via the IPTP model, the PSAST model is then employed to simulate the aqueous tracer transport in the varied aqueous pore space. The simulation flowchart is given in Fig. 6.4 to illustrate the simulation strategy.

It should be noted that the immiscible interface between TCE and water can absorb aqueous tracer due to the physical and chemical properties of tracer. However, we focus on the influence of water saturation topology on the aqueous-miscible tracer transport. The surface reaction on immiscible interface is neglected. Thus, the non-reactive aqueous-miscible tracer is considered and the immiscible interfaces between TCE, water and soil are assumed as a non-slip boundary for the aqueous tracer transport. The no-slip boundary condition is applied at the solid–liquid and liquid–liquid interfaces. There is no flux across the top and bottom boundaries. For the aqueous solute transport, the pore space is initially occupied by water, and there is no tracer in the aqueous pore space. The pulsed tracer with concentration C_0 is injected along the inlet boundary. The step function is used to simulate the pulsed injection condition. Figure 6.5 shows that the distribution of residual NAPL and flow field in vaired pore porous media and it can be seen that the aqueous flow pathways were significantly changed due to the residual TCE.

A dimensionless parameter (i.e., pore volume, P_v) related to the time is introduced as follows,

$$P_\text{v} = Qt/V \tag{6.4.1}$$

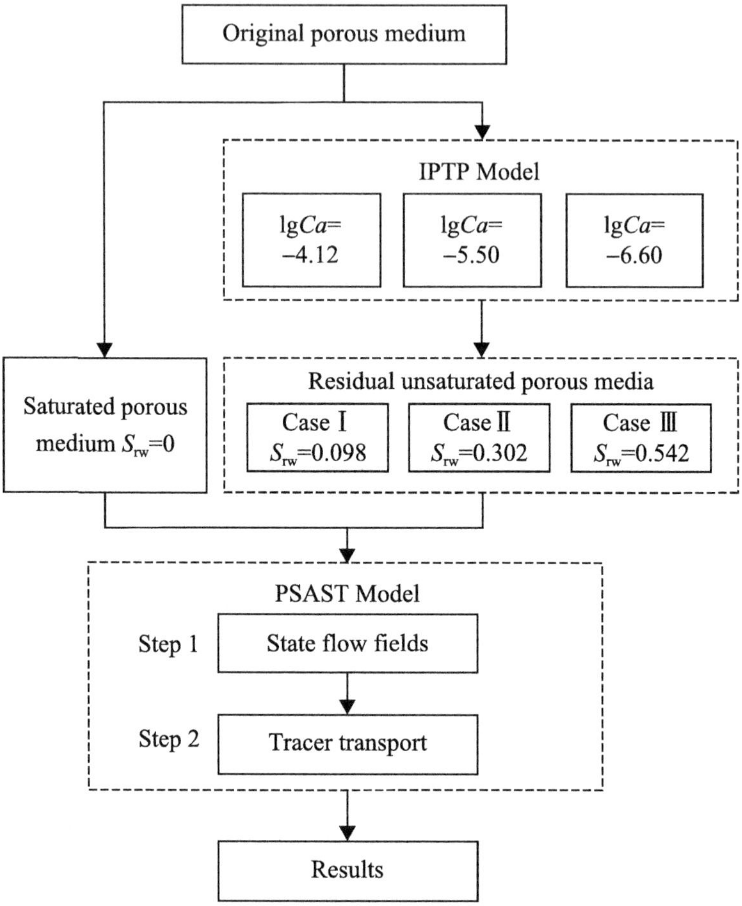

Fig. 6.4 Flowchart of simulation

where Q is the flow rate [L^3T^{-1}] per length and V is the total volume of the pore space domain [L^3]. A solute is introduced by a pulse injection method, with a flux weight distribution along the inlet boundary. The whole space in the porous medium is initially saturated with pure water. A total of 0.05 P_v of a solute, with an initial concentration C_0 of 1 mol/m^3, is injected.

Flow and solute transport simulations are performed in pore-scale regions by imposing pressure gradients resulting in certain dimensionless Peclet number (*Pe*) defined by

$$Pe = \tau_D/\tau_a = 2u_{av}\overline{r}/D_m \tag{6.4.2}$$

which is the ratio of the characteristic diffusion time [$\tau_D = (2\overline{r})^2/D_m$] to the characteristic advection time ($\tau_a = 2\overline{r}/u_{av}$), where u_{av} is the average velocity magnitude

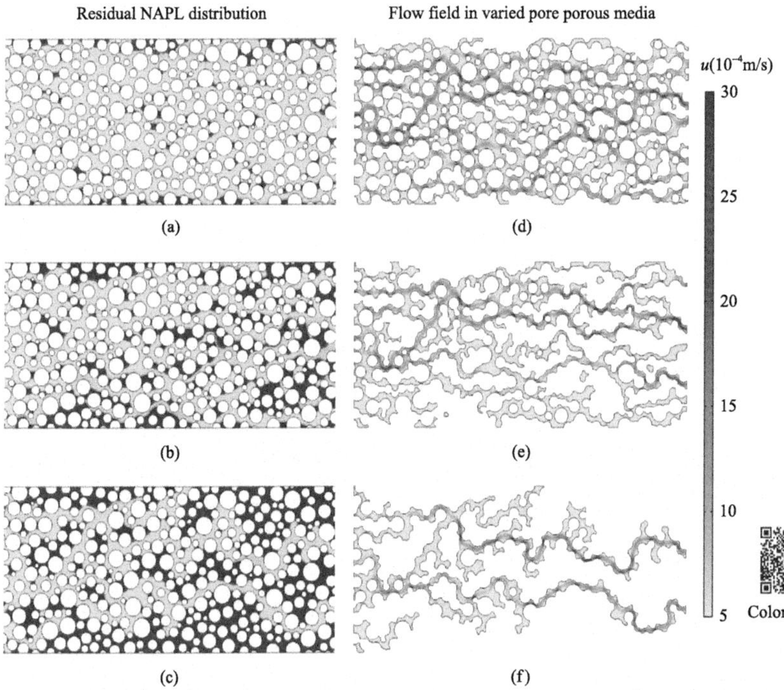

Color picture

Fig. 6.5 Results from coupling IPTP model and PSAST model **a** S_{rw} = 0.09778, lgCa = –4.117; **b** S_{rw} = 0.30212, lgCa = –5.5033; **c** S_{rw} = 0.54244, lgCa = –6.6019; **d** S_{rw} = 0.09778, lgCa = –4.117; **e** S_{rw} = 0.09778, lgCa = –4.117; **f** S_{rw} = 0.09778, lgCa = –4.117

in the porous media and $\overline{r}$ is the average radius of the grains. Pe = 10 and 100 are considered for PM0-3 corresponding to the average velocity magnitude 0.481 m/d and 4.806 m/d in PM0, respectively, representing the common natural flow and high-speed flow conditions.

6.5 Behaviors of Aqueous Tracer Mass Transfer

Figure 6.6 shows the aqueous tracer transport in both the original porous media and Case I–III. Due to the significant change of the pore structure caused by the residual TCE, the spatial evolution of the aqueous tracer in Case I–III is significantly different from that in the original aqueous flow field. As the residual TCE saturation increases, the area of the potential zone for aqueous tracer transport decreases. For the capillary fingering displacement (e.g., Case III), the aqueous tracer pathways form several fingers and are independent of the pore structure in original porous media. Several dead-end pathways for aqueous tracer transport are observed in both Case II and Case III. The aqueous tracer transport into these dead-end pathways is limited due to the slow diffusion-dominant process. It should be mentioned that the values of *Pe* are greater than 200 in all cases (see Table 6.1), indicating that the aqueous tracer transport is advection-dominant.

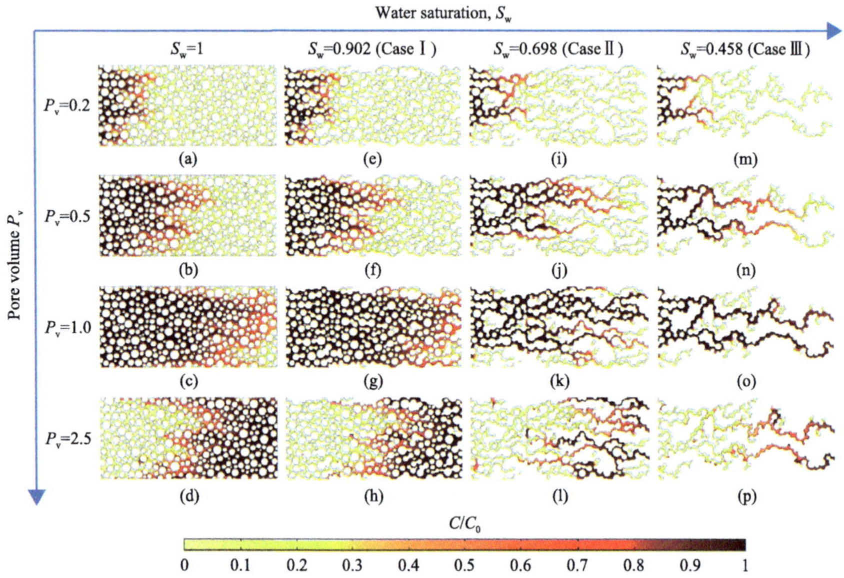

Fig. 6.6 The aqueous tracer transport in both the original aqueous porous media and Case I–III C/C_0 is the dimensionless concentration

Table 6.1 Parameters of the aqueous flow fields in the original porous media and Case I–III

Porous media	$\overline{u_x}$(m/s)	$\overline{u}$(m/s)	$\sigma^2(u)$(m^2/s^2)	σ^2ln$u$	*Pe*	*Re*	CV_u
Original (S_w = 1.000)	4.93 × 10–^{4}	6.30 × 10–^{4}	3.47 × 10–^{7}	3.79	393.87	0.39	0.89
Case I (S_w = 0.902)	5.26 × 10–^{4}	6.65 × 10–^{4}	3.45 × 10–^{4}	7.15	415.59	0.41	0.86
Case II (S_w = 0.698)	3.60 × 10–^{4}	4.61 × 10–^{4}	2.55 × 10–^{4}	58.40	287.97	0.29	1.19
Case III (S_w = 0.458)	3.25 × 10–^{4}	4.59 × 10–^{4}	2.03 × 10–^{4}	131.19	286.93	0.29	1.24

Note $\overline{u_x}$ is the mean longitudinal flow velocity; $\sigma^2(u)$ is the variance of the flow velocity; σ^2ln$u$ is the natural logarithm of $\sigma^2(u)$; *Pe* is dimensionless Peclet number; *Re* is dimensionless Reynolds number

However, the aqueous tracer transport in several dead-end pathways is diffusion-dominant due to the extremely low flow velocity. As is shown in Figs. 6.6l, p, once the aqueous tracer is transported into the dead-end pathways, a certain amount of aqueous tracer is entrained and the mass exchange processes between the dead-end pathways and the preferential pathways become rate-limited. Consequently, the trapped aqueous tracer in dead-end pathways can lead to anomalous transport behaviors (e.g., long tails of the BTCs). In addition, due to the enhanced velocity spatial variability, the aqueous tracer in the preferential flow pathways not only is transported faster but also has a higher concentration in Case II and Case III than in the original porous media and Case I.

As a cumulative consequence of aqueous tracer transport through the unsaturated porous media system, the BTCs reflect the temporal evolution of the aqueous tracer. In Fig. 6.7, all BTCs exhibit typical characteristics of anomalous aqueous tracer transport (e.g., the early arrival and long tails of the BTCs), although the aqueous tracer is transported in the original porous media where no residual TCE changes the pore structure. This indicates that not only the residual TCE but also the heterogeneity of the original porous media can lead to the anomalous aqueous tracer transport.

As is shown in Fig. 6.7, the early arrival and long tails of the BTCs are the same in the original porous media and Case I. This is because that the aqueous flow field is almost the same, and the compact displacement pattern in Case I efficiently displaces the defending TCE from the pore space and the variation of pore structure is trivial. As the capillary number decreases, the immiscible fluid–fluid displacement pattern markes the transition to the capillary fingering regime and the residual TCE saturation increases, associating the significantly varied pore structure. Thus, the velocity spatial variability in the aqueous flow field is enhanced. Consequently, the early arrival and long tails of the BTCs are pronounced in Case II and Case III than in

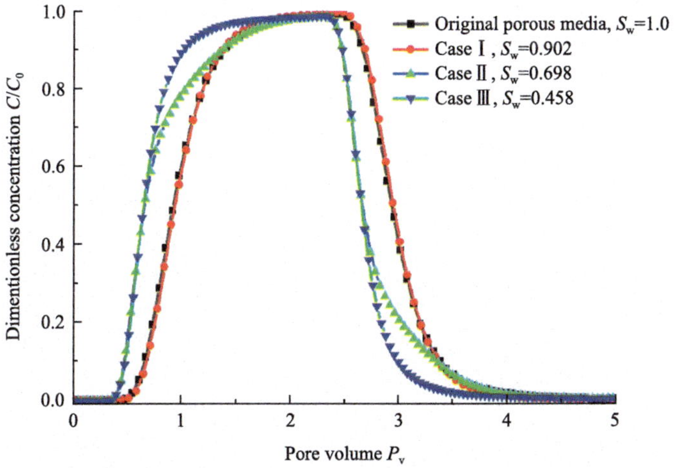

Fig. 6.7 The breakthrough curves (BTCs) of the aqueous tracer transport in both the original aqueous porous media and Case I–III

the original porous media and Case I. Although the residual TCE significantly varies the pore structure in Case II and Case III, the early arrival of the BTCs is the same while the long tails of the BTCs are different in Case II and Case III. This indicates that the residual TCE in Case II and Case II has few influences on the mean flow velocity but has significant influences on the mass transfer between the mobile and immobile domains.

To evaluate quantitatively the anomalous transport characteristics, we use both traditional ADE model and CTRW model inversely to fit the BTCs. Detailed information on the goodness-of-fitting of both ADE model and CTRW model can be found in the Supporting Information. Table 6.2 shows the goodness-of-fit results of the ADE and CTRW models.

As listed in Table 6.2, the coefficient of determination R^2 is about 1.0 and the global error E_i is low, indicating the ability to characterize aqueous tracer transport by using both ADE and CTRW model. As a key parameter in the CTRW model, β_{CTRW}

Table 6.2 Results from the goodness-of-fitting of the ADE and CTRW models

Porous media	ADE model					CTRW model	
	$\overline{u_{ADE}}$ (m/s)	D_{ADE} (m²/s)	$\alpha_{L,ADE}$ (m)	E_i	R^2	$\overline{u_{CTRW}}$ (m/s)	D_{CTRW}(m²/s)
Original (S_w = 1.000)	5.15×10^{-4}	9.40×10^{-7}	1.83×10^{-3}	0.009	0.999	5.17×10^{-4}	9.37×10^{-7}
Case I (S_w = 0.902)	5.42×10^{-4}	8.61×10^{-7}	1.59×10^{-3}	0.006	1.000	5.48×10^{-4}	8.43×10^{-7}
Case II (S_w = 0.698)	4.78×10^{-4}	1.94×10^{-7}	4.06×10^{-3}	0.030	0.994	4.82×10^{-4}	1.96×10^{-7}
Case III (S_w = 0.458)	4.74×10^{-4}	8.64×10^{-7}	1.82×10^{-3}	0.018	0.998	4.76×10^{-4}	8.72×10^{-7}
Porous media	CTRW model						
	$\alpha_{L,CTRW}$(m)	T_1(s)	T_2(s)	β_{CTRW}	E_i	R^2	
Original (S_w = 1.000)	1.81×10^{-3}	0.01	1405.72	1.91	0.016	0.999	
Case I (S_w = 0.902)	1.54×10^{-3}	0.01	1078.95	1.92	0.014	1.000	
Case II (S_w = 0.698)	4.07×10^{-3}	0.01	1120.73	1.80	0.030	0.990	
Case III (S_w = 0.458)	1.83×10^{-3}	0.01	968.95	1.90	0.023	0.998	

characterizes the dispersion regime. According to the previous study (Cortis and Berkowitz 2004), $0 < \beta_{CTRW} < 1$ indicates a highly anomalous transport; $1 < \beta_{CTRW} < 2$ is associated with a moderately anomalous transport; $\beta_{CTRW} > 2$ correspondes to the Fickian transport. Table 6.2 shows that the values of all β_{CTRW} are close to 2.0, indicating that all aqueous tracer transports in this study are moderately anomalous. The mean flow velocities $\overline{u_{ADE}}$ and $\overline{u_{CTRW}}$ are dependent on the residual TCE distribution and show the same trend as the pore structure varies by the residual TCE. The dispersivities are determined as $\alpha_{L,ADE}(= D_{ADE}/\overline{u_{ADE}})$ and $\alpha_{L,CTRW} = (D_{CTRW}/\overline{u_{CTRW}})$ in ADE and CTRW models, respectively. In Table 6.2, the results indicate a non-monotonic relationship between the dispersivity and S_w in both ADE and CTRW models. Due to the high S_w and almost unchanged pore structure, the dispersivity is slightly lower in Case II than in the original porous media. This suggests that the residual TCE resulting from the compact fluid–fluid displacement process has a small influence on the aqueous tracer transport. It is found that the dispersivity is enhanced in Case II when the pore structure is significantly changed and S_w is decreased due to the residual TCE. However, the dispersivity shows a slight decrease in Case III where the pore structure is significantly changed and the velocity spatial variability of the aqueous flow field is enhanced. It should be noted that since both ADE and CTRW models are assumed to be one-dimensional, the dispersivity by both ADE and CTRW models can be expected to be the averaged dispersivity over the whole domain. Therefore, it is insufficient to characterize the mass transfer of the aqueous tracer transport only by the dispersivity. In other words, our results indicate that the dispersivity of aqueous tracer is dependent on not only the S_w but also the distribution of the residual TCE related to the immiscible fluid–fluid displacement regime.

Since the residual TCE changes the pore structure and space of the wetting phase which provides the potential space for the aqueous flow field and the aqueous tracer transport, the corresponding flow field is consequently split into flowing and stagnant regions, referred to as mobile and immobile regions for aqueous tracer transport. The solute exchange between the mobile and immobile regions can substantially delay the transport of the aqueous tracer, associated with the tailing of the tracer BTCs. The tailing of the tracer BTCs is also related to the reaction at the solid–fluid interface, which can be assessed by the Sherwood number. In this study, the tracer is assumed to be non-reactive. Therefore, we focus on the solute exchange between the mobile and immobile regions. To quantify this mechanism, we propose to estimate the mass transfer for aqueous tracer transport by MIM model. The MIM model is a time nonlocal transport model, which describes the diffusion-driven mass exchange between the mobile and immobile domains. The single mass-transfer rate MIM model can capture the non-Fickian solute transport caused by the diffusion-limited transport of solutes in immobile zones. The MIM transport equation is given by (van Genuchten and Wierenga 1976)

$$\frac{\partial C_m}{\partial t} + \left(\frac{S_{sw}}{S_w - S_{sw}}\right)\frac{\partial C_{im}}{\partial t} = -\overline{u_m}\frac{\partial C_m}{\partial x} + D_{f,mim}\frac{\partial^2 C_m}{\partial x^2} \tag{6.5.1}$$

where C_m and C_{im} represent the concentration of the aqueous tracer in the mobile (flowing) and immobile (stagnant) domains in the wetting phase respectively; $D_{f,mim}$ is the fitted dispersion coefficient and $\overline{u_m}$ is the effective flow velocity in mobile domain for the wetting phase; S_w is the saturation of wetting phase, which is equal to the sum of the stagnant saturation S_{sw} and the flowing saturation S_{fw} in the wetting phase. A partitioning coefficient $\beta = (S_w - S_{sw})/S_w = \frac{S_{fw}}{S_w}$ is introduced as the mobile domain fraction in the wetting phase. Thus, the $\overline{u_m}$ in Eq. (6.5.1) is related to the mean flow velocity of the wetting phase ($\overline{u_{MIM}}$) by $\overline{u_m} = \overline{u_{MIM}}/\beta$. The dispersity $\alpha_{L,mim}(= D_{f,mim}/\overline{u_m})$ in MIM model is determined for all cases. The solute exchange between the mobile and immobile domains is described by a nonequilibrium mass transfer model,

$$S_{sw}\frac{\partial C_{im}}{\partial t} = \alpha(C_m - C_{im}) \tag{6.5.2}$$

where α is the mass transfer coefficient between the mobile and immobile domains. For MIM model, the best-fit parameters $D_{f,mim}$, α, and β are estimated. The initial estimate of the parameters has a great influence on the accuracy of the inverse parameters, and an accurate initial estimation can improve the convergence to a minimum error. The initial values of $\overline{u_m}$ and $D_{f,mim}$ are set to the counterpart values calculated by the ADE model. The STANMOD V2.08 based on CXTFIT code (Toride et al. 1995) is employed for the inversion of MIM model.

Table 6.3 lists the parameters estimated from MIM model. As is shown in Table 6.3, both the R^2 close to 1.0 and the low E_i ensure the reliability and accuracy of the inversion of MIM model. As is listed in Table 6.3, β shows a non-monotonic decrease with decreasing S_w. Note that our result of β reflects two aspects. First, the maximum β is observed at $S_w = 1.0$ (e.g., saturated conditions). Second, the low β is observed at $S_w = 0.698$ in Case II rather than at lower saturation (e.g., $S_w = 0.458$ in Case III). This implies that the larger domain fraction in the wetting phase become immobile for aqueous tracer transport at $S_w = 0.458$, rather than at $S_w = 0.698$. This can be attributed to the discontinuous residual TCE formed by the different patterns of the immiscible fluid–fluid displacement. More especially, the residual distribution of TCE at $S_w = 0.458$ results from the transition of the displacement pattern from the compact displacement to the capillary fingering regime. Unlike compact displacement and capillary fingering displacement, there are remarkable dead-end pathways with the stagnant flow velocity in this transition displacement, which is consequently unfavorable for advection- dominant transport of the aqueous tracer. Therefore, β in residual unsaturated porous media is affected by not only S_w but also the residual saturation topology.

In Table 6.3, the maximum $\alpha_{L,mim}$ is observed at $S_w = 0.698$ in Case II, and the relationship between $\alpha_{L,mim}$ and S_w shows a non-monotonic relation, for both ADE and CTRW models. It should be mentioned that $\alpha_{L,mim}$ is significantly smaller than both $\alpha_{L,ADE}$ and $\alpha_{L,CTRW}$, since $\alpha_{L,mim}$ accounts only for the dispersivity of the mobile domain in the wetting phase. According to the dispersion regime of MIM model, an overall dispersivity includes both the dispersivity of the mobile domain

Table 6.3 Summary of the parameters estimated from the MIM model

Porous media	$\overline{u_m}$ (m/s)	$D_{f,min}(m^2/s)$	β	$\alpha(d^{-1})$	$\alpha_{L,min}(m)$	R^2	E_i
Original (S_w = 1.000)	5.55×10^{-4}	7.07×10^{-7}	0.93	1.63×10^{-3}	1.99×10^{-3}	0.999	0.006
Case I (S_w = 0.902)	5.89×10^{-4}	6.59×10^{-7}	0.92	2.36×10^{-3}	1.67×10^{-3}	0.999	0.005
Case II (S_w = 0.698)	6.50×10^{-4}	4.42×10^{-7}	0.74	2.59×10^{-3}	4.45×10^{-3}	0.998	0.018
Case III (S_w = 0.458)	5.28×10^{-4}	5.36×10^{-7}	0.90	5.10×10^{-4}	2.92×10^{-3}	0.999	0.016

and the influence of solute exchange between the mobile and immobile domains. Thus, the overall dispersivity α_{MIM} in terms of the MIM parameters can be defined (Toride et al. 2003)

$$\alpha_{MIM} = \alpha_{L,mim} + \frac{S_{fw}^2 S_{sw} \overline{u_m}}{\alpha S_w^2} \tag{6.5.3}$$

We introduce $\alpha_{L,ms}$ to represent the contribution of the solute exchange between the mobile and immobile domains to the overall dispersivity α_{MIM}.

$$\alpha_{L,ms} = \frac{S_{fw}^2 S_{sw} \overline{u_m}}{\alpha S_w^2} \tag{6.5.4}$$

The dispersivities from the MIM and ADE models as functions of water saturation are shown in Fig. 6.8. $\alpha_{L,ADE}$ from the ADE model is always smaller than the overall dispersivity α_{MIM} from the MIM model over all S_w, which is consistent with previous studies (Toride et al. 2003; Zhuang et al. 2021). This can be due to the inherently limited capability of ADE model for characterizing the anomalous BTCs. On the other hand, this is reasonable to characterize the overall dispersivity into the two mass transfer mechanisms (e.g., dispersion in mobile domain and solute exchange between immobile and mobile domain). As a result, α_{MIM} as the overall dispersivity is always larger than both $\alpha_{L,ms}$ and $\alpha_{L,mim}$ over all S_w. Moreover, the non-zero of $\alpha_{L,ms}$ and $\alpha_{L,mim}$ confirms the coexistence of these two mass transfer mechanisms over all S_w.

It should be mentioned that the relative contribution of these two mass transfer mechanisms to the overall dispersivity is of great importance to understanding aqueous tracer transport behavior. Therefore, to evaluate the relative contribution between $\alpha_{L,ms}$ and $\alpha_{L,mim}$ in the dispersion regime of MIM model,

$$\varepsilon_d = \frac{\alpha_{L,ms}}{\alpha_{L,mim}} \tag{6.5.5}$$

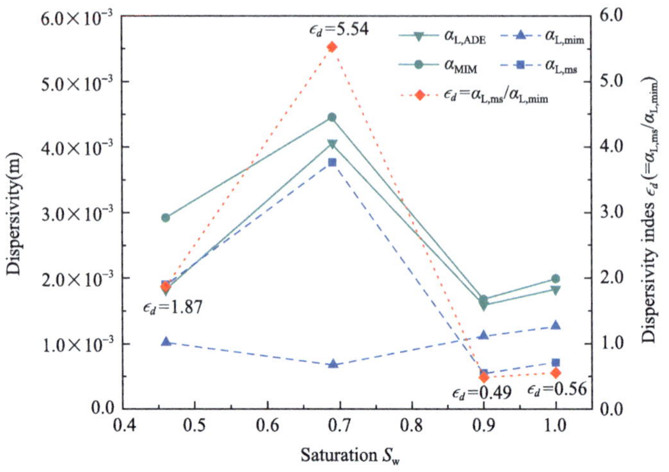

Fig. 6.8 Comparison the dispersivities from MIM and ADE models The overall dispersivity α_{MIM} was calculated with Eq. (6.5.3), the $\alpha_{L,ms}$ was determined by Eq. (6.5.4), the dimensionless dispersivity index ε_d was obtained by Eq. (6.5.5)

$\varepsilon_d > 1$ means that the overall dispersivity is dominated by the solute exchange between the mobile and immobile domains, whereas $\varepsilon_d < 1$ indicates that the overall dispersivity is controlled by the dispersion of aqueous tracer in the mobile domain. As is shown in Fig. 6.8, ε_d as a function of S_w shows a non-monotonic relationship, since the similar relationship is observed between α_{MIM} and S_w. The values of ε_d are 0.56, 0.49, 5.54, and 1.87 at $S_w = 1.0, 0.902, 0.698$, and 0.458, respectively, indicating that the mass transfer of aqueous at high water saturation conditions (e.g., $S_w = 1.0$ and 0.902) is controlled by the dispersion of aqueous tracer in the mobile domain, while at low water saturation conditions (e.g., $S_w = 0.698$ and 0.458) it is dominated by solute exchange between the mobile and immobile domains. This suggests that the mass transfer mechanism of aqueous tracer varies with S_w. Interestingly, the mass transfer coefficient between the mobile and immobile domains (see α in Table 6.3) is found to have an overall decreasing trend at $\varepsilon_d > 1$. Note that $1/\alpha$ represents the time scale for solute exchange between the mobile and immobile domains. $1/\alpha$ is three times larger under water-saturated conditions than under conditions of low water saturation conditions. This means that the exchange time was longer at low S_w, which is related to slower solute mixing. Thus, if mass transfer of the aqueous tracer is dominated by solute exchange between the mobile and immobile domains at low S_w, the long exchange time will cause aqueous tracer transport to be retained and mass transfer to be delayed. In addition, the proposed ε_d will provide insights into the anomalous characteristics of mass transfer of aqueous tracer in residual unsaturated porous media. Recall the anomalous characteristics of BTCs in Fig. 6.7: The dominant solute exchange between the mobile and immobile domains causes the remarkable long tails of BTCs at $\varepsilon_d > 1$, while the early arrival of BTCs is observed at $\varepsilon_d < 1$, where dispersion in the mobile domain controls the mass transfer of the aqueous tracer.

References

Badalassi VE, Ceniceros HD, Banerjee S (2003) Computation of multiphase systems with phase field models. J Comput Phys 190(2):371–397

Benzi R, Biferale L, Sbragaglia M et al (2006) Mesoscopic modeling of a two-phase flow in the presence of boundaries: the contact angle. Phys Rev E 74(2). https://doi.org/10.1103/PhysRevE.74.021509

Cortis A, Berkowitz B (2004) Anomalous transport in "classical" soil and sand columns. Soil Sci Soc Am J 68(5):1539–1548

Gunstensen AK, Rothman DH (1991) Lattice Boltzmann model of immiscible fluids. Phys Rev A 43(8):4320–4327

He X, Chen S, Zhang R (1999) A Lattice Boltzmann scheme for incompressible multiphase flow and its application in simulation of Rayleigh-Taylor Instability. J Comput Phys 152:642–663

Huang H, Lu X-Y (2009) Relative permeabilities and coupling effects in steady-state gas-liquid flow in porous media: a lattice Boltzmann study. Phys Fluids 21(9):092104

Hysing S, Turek S, Kuzmin D et al (2009) Quantitative benchmark computations of two-dimensional bubble dynamics. Int J Numer Meth Fluids 60(11):1259–1288

Jacqmin D (1999) Calculation of two-phase Navier-Stokes flows using phase-field modeling. J Comput Phys 155(1):96–127

Kang Q, Zhang D, Chen S (2002) Displacement of a two-dimensional immiscible droplet in a channel. Phys Fluids 14(9):3203

Lenormand R, Touboul E, Zarcone C (1988) Numerical models and experiments on immiscible displacements in porous media. J Fluid Mech 189(1):165–187

Martys NS, Chen H (1996) Simulation of multicomponent fluids in complex three-dimensional geometries by the lattice Boltzmann method. Phys Rev E 53(1):743–751

Qian YH, D'Humières D, Lallemand P (1992) Lattice BGK models for Navier-Stokes equation. Europhys Lett 17(6):479–484

Shan X, Chen H (1993) Lattice Boltzmann model for simulating flows with multiple phases and components. Phys Rev E 47(3):1815–1820

Shan X, Chen H (1994) Simulation of nonideal gases and liquid-gas phase transitions by the lattice Boltzmann equation. Phys Rev E 49(4):2491–2498

Shan X, Doolen DG (1994) Multicomponent lattice-Boltzmann model with interparticle interaction. J Stat Phys 81(1–2):379–393

Swift MR, Osborn WR, Yeomans JM (1995) Lattice Boltzmann simulation of nonideal fluids. Phys Rev Lett 75(5):830–833

Swift MR, Orlandini E, Osborn WR et al (1996) Lattice Boltzmann simulations of liquid-gas and binary fluid systems. Phys Rev E 54(5):5041–5052

Toride N, Inoue M, Leij FJ (2003) Hydrodynamic dispersion in an unsaturated dune sand. Soil Sci Soc Am J 67(3):703–712

Toride N, Leij F, van Genuchten MT (1995) The CXTFIT code for estimating transport parameters from laboratory or filed tracer experiments. US Salinity Laboratory Riverside, Riverside

van Genuchten MT, Wierenga P (1976) Mass transfer studies in sorbing porous media I. Analytical solutions. Soil Sci Soc Am J 40(4):473–480

Villanueva W, Amberg G (2006) Some generic capillary-driven flows. Int J Multiph Flow 32(9):1072–1086

Yue P, Feng JJ, Liu C et al (2004) A diffuse-interface method for simulating two-phase flows of complex fluids. J Fluid Mech 515:293–317

Zhuang L, Raoof A, Mahmoodlu MG et al (2021) Unsaturated flow effects on solute transport in porous media. J Hydrol 598:126301

Chapter 7
Mass Transfer Between Matrix and Filled Fracture During Imbibition Process

The spontaneous imbibition (SI) is the spontaneous entry of a wetting fluid into a porous medium driven only by capillary forces, displacing a preexisting nonwetting fluid. As a fundamental mass transfer process, the importance of SI mechanism has long been recognized in many hydrological and geotechnical applications, such as sequestration of carbon dioxide, oil and gas extraction, and the protection of groundwater resources. This chapter focuses on the mass transfer phenomena of the imbibed water between the matrix and the filled fracture during the SI process based on the LF-NMR technology.

7.1 LF-NMR Measurement and Principle

Quantitively distinguishing the origin of the imbibed water from the matrix and the fracture is crucial for characterizing the mass transfer between the matrix and the fracture. In recent years, many advanced technology has improved the experimental capability for analyzing the SI process, such as neutron radiographic techniques (Cheng et al. 2015; Zhao et al. 2019), CT scanning techniques (Akin and Kovscek 1999; Alyafei et al. 2016; Zhao et al. 2017), low-field nuclear magnetic resonance technique (LF-NMR) (Meng et al. 2015; Lyu et al. 2019; Yuan et al. 2019). Due to its rapid, non-invasive, and quantitative advantages, the LF-NMR has won worldwide recognition and has demonstrated a reliable ability to quantitatively characterize pore structure and seepage behavior. For the SI process, the LF-NMR can be used to distinguish the amount of water imbibed in pores of different sizes, and it is capable of characterizing the type of the water content (Meng et al. 2015; Yuan et al. 2019; Chen et al. 2021). Since the SI process and the mass transfer between the matrix and the fracture strongly depend on the multi-scale pore system and the complex pore structure, the SI experiment in a dual-porosity media with the porous matrix and the filled fracture is conducted based on the LF-NMR.

Z. Dou et al., *Mass Transfer Dynamics of Contaminants in Fractured Media*,
https://doi.org/10.1007/978-981-99-9187-7_7

The standard Carr-Purcell-Meiboom-Gill (CPMG) pulse sequence is used to measure the transverse relaxation time (T_2) during the LF-NMR measurement. This CPMG pulse sequence provides 200 data points in one echo acquisition time. The number of echoes acquired by each CPMG sequence is 10,000. The corresponding CPMG parameters are set as follows: An echo spacing (T_E) of 0.15 ms, a waiting time (T_W) of 6000 ms, an echo number of 2048, and the scanning number (N_S) of 16.

The NMR signal or magnetization (M_t) from a fluid containing non-zero spin state (i.e., ^1H) nuclei that is bound in a porous medium is the convolution of all contribution volumes related to n pores. The signal attenuation is characterized by a different T_2 relaxation constant. The M_t can be described as follows (Connolly et al. 2017):

$$\frac{M_t}{M_0} = \frac{1}{V_0}\sum_{i}^{n}\left(V_i \times e^{\frac{-t}{T_{2,i}}}\right) \tag{7.1.1}$$

where V_0 and V_i are the total volume of sample and the ith pore volume contribution, respectively. M_0 is the equilibrium magnetization and $T_{2,i}$ is the relaxation rate for the ith pore.

The NMR transverse relaxation spectroscopy technique is used during the LF-NMR measurement. The typical NMR relaxation process includes bulk relaxation (T_{2b}), surface relaxation (T_{2s}) and diffusion relaxation (T_{2d}) (Coates et al. 1999). The T_2 is controlled by these three different relaxations and can be expressed as

$$\frac{1}{T_2} = \frac{1}{T_{2b}} + \frac{1}{T_{2s}} + \frac{1}{T_{2d}} \tag{7.1.2}$$

$$T_{2b} \approx 3\left(\frac{T_k}{298\mu}\right) \tag{7.1.3}$$

$$\frac{1}{T_{2s}} = \rho_2\left(\frac{S}{V}\right)_{pore} \tag{7.1.4}$$

$$\frac{1}{T_{2d}} = \frac{D(\gamma G T_E)^2}{12} \tag{7.1.5}$$

The T_{2b} is inversely proportional to the fluid viscosity. Since the viscosity of water is small, the $1/T_{2b}$ is small enough to be ignored (Coates et al. 1999). T_{2s} is the lateral relaxation time caused by surface relaxation and is inversely proportional to the self-diffusion coefficient of liquids and the square of the echo spacing T_E (Bendel 1990). The $1/T_{2d}$ is negligible when the T_E is small, since the diffusion of water molecules is very fast. Consequently, the T_2 is dominated by the T_{2s} and Eq. (7.1.2) can be simplified as (Wang et al. 2010; Chen et al. 2021; Hu et al. 2021):

$$\frac{1}{T_2} = \rho_2 \left(\frac{S}{V} \right)_{\text{pore}} \approx \rho_2 \frac{2}{R} \tag{7.1.6}$$

where ρ_2 is the surface relaxation ratio and R is the radius of the pore.

From Eq. (7.1.6), it can be seen that T_2 is proportional to the pore radius, which means that the larger the pore radius, the stronger the NMR signal. In fact, T_2 distribution reflects the pore size distribution when the sample is saturated. Therefore, the water imbibed into the glass bead-filled fracture or the matrix of the yellow sandstone can be easily distinguished during the SI processes because the distribution of the pore space between the glass bead-filled fracture and the matrix of the sample is clearly different with the resulting T_2 distribution being different. This provides a reliable non-invasive method to study the mass transfer behavior between the glass bead-filled fracture and the matrix of the sample during the SI processes.

7.2 Experimental Materials

Since yellow sandstone is widespread in nature and is a typical reservoir rock, yellow sandstone cores are selected as matrix in dual-porosity media for the SI experiment. The sandstone cores are cylinders with a diameter of about 2.4 cm and a length of about 5.9 cm. Some of the cores are drilled a hole in the middle to mimic the fracture in the matrix [e.g., Sample S in Fig. 7.1b]. To further structure a matrix-filled fracture system, the hole in the core is filled with glass beads and under this condition the core sample is considered as a dual-porosity medium [e.g., Sample D_1 in Fig. 7.1c].

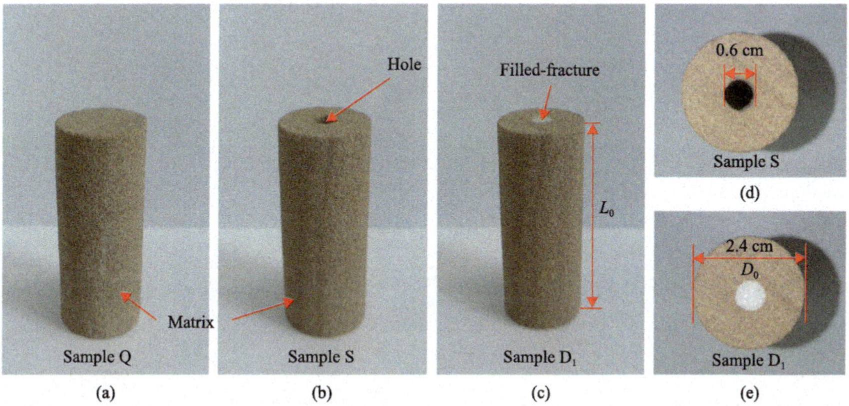

Fig. 7.1 Different yellow sandstone cores as single-porosity and dual-porosity media. **a** Sample Q: The yellow sandstone core without a hole; **b** Sample S: The yellow sandstone cores with an unfilled hole (seen as the single-porosity media); **c** Sample D_1: The yellow sandstone cores with a glass bead-filled fracture (seen as dual-porosity media); **d, e** are vertical views of the Sample S and Sample D_1.

The main component of glass beads is silica, which shows no signal in the NMR test. Therefore, the composition and content of glass beads do not affect the experimental results. And when used as fracture fillers, the glass beads are loose enough to be well distinguished from the yellow sandstone in the matrix. Five different sizes of glass beads with the same mass are uniformly packed into the hole of sample S to ensure consistency of porosity. As is shown in Table 7.1, the porosity of the matrix.

in all samples is 12.24%. The particle size of the filled glass beads varies significantly between samples D_1–D_5, while the porosity of the glass bead-filled fracture of all samples is almost the same, reaching about 33.32%. Although the porosity in the glass bead-filled fracture is only nearly three times that in the matrix of the yellow sandstone core, the pore structure and connectivity between the glass bead-filled fracture and the matrix differ significantly due to the different formation patterns. Therefore, the yellow sandstone with glass bead-filled fracture can be considered as a matrix-filled fracture system of dual-porosity media for the SI process.

The porosity of the matrix of all samples is determined based on the LF-NMR measurement. The porosity of the glass bead-filled fracture of all samples is determined based on the dry unit weight of the filled glass beads which is measured by the hydrometer method (Das 2019). In addition, ultrapure water is used in all SI experiments at a room temperature of 25°C.

Table 7.1 Physical properties of all samples with single and dual-porosity media

Media types	SampleID	L_0 (mm)	D_0 (mm)	Glass beads			Φ_1 (%)	Φ_2 (%)
				Particle size (mm)	Dry unit weight (g/cm^3)	Mass (g)		
Single-porosity media	Q	59.80	24.50				10.53	
Single-porosity media	S	59.60	24.20				12.24	
Dual-porosity media	D_1	59.60	24.20	0.071–0.090	2.402	2.781	12.24	31.29
Dual-porosity media	D_2	59.60	24.20	0.125–0.180	2.479	2.781	12.24	33.43
Dual-porosity media	D_3	59.60	24.20	0.280–0.355	2.489	2.781	12.24	33.70
Dual-porosity media	D_4	59.60	24.20	0.355–0.500	2.493	2.781	12.24	33.80
Dual-porosity media	D_5	59.60	24.20	0.500–0.710	2.514	2.781	12.24	34.36

Note L_0 and D_0 are the length and diameter of the sample, respectively. Φ_1 and Φ_2 are the porosity of the matrix and the filled fracture, respectively

7.3 Distribution of the Imbibed Water

The T_2 spectrum during SI processes for sample S and samples D_1–D_5 is compared in Fig. 7.2. It can be seen that the T_2 spectrum of sample S (see Fig. 7.2a) is a unimodal distribution, whereas the T_2 spectrum of samples D_1–D_5[see Fig. 7.2b–f] shows a bimodal distribution. This difference between samples S and D_1–D_5 can be attributed to the different distribution of the pore space. It should be mentioned that the T_2 relaxation time is proportional to the pore space when the pore is saturated. The distribution of the pore space in sample S is determined by the matrix and is concentrated in a relatively narrow range, whereas the distribution of the pore space in samples D_1–D_5 are determined by both the matrix and the glass beads. Since there is a multi-scale pore system between the matrix and the glass bead-filled fracture in samples D_1–D_5, the T_2 spectrum of samples D_1–D_5 are bimodal, while the T_2 spectrum of sample S is unimodal. This result also indicates that the samples D_1–D_5 are not only a dual-porosity media but also has two different distributions of pore space between the matrix and the glass bead-filled fracture. Furthermore, it is found that as the range of the particle size of the glass beads increases from 0.071–0.090 to 0.500–0.710 mm, the peak of T_2 spectrum varies from about 100–800 ms for samples D_1–D_5. This variation in the peak of the T_2 spectrum is caused by the increasing particle size of the glass beads.

Based on the multi-scale pore system and the resulting distinct T_2 spectrum in samples D_1–D_5, we can quantitatively distinguish the distribution of the imbibed water from the matrix and glass bead-filled fracture. Here, we introduce a T_2 cut-off method to identify the distribution of the imbibed water in the matrix or the glass bead-filled fracture. To determine the cut-off value of the T_2 spectrum in the matrix and the glass bead-filled fracture, the referencing sample S is saturated. Then the T_2 spectrum of the saturated sample S is measured using LF-NMR. It should be noted that the sample S has the same matrix as samples D_1–D_5, but the fracture is not filled with the glass beads. Comparing the saturated T_2 spectrum in samples S and D_1–D_5 in Fig. 7.3, we find that the T_2 distributions in both samples S and D_1–D_5 almost overlap when T_2 is less than 16 ms. Since the T_2 distribution in sample S only reflects the water distribution in the matrix, the overlapping T_2 distribution in samples D_1–D_5 can be derived as a T_2 spectrum from the water distribution in the same matrix.

It is also evident from Fig. 7.3 that T_2 does not overlap for more than 16 ms and varies with samples D_1–D_5. This can be attributed to the water distributions in the different glass bead-filled fracture in samples D_1–D_5. It should be mentioned that the integrated area of the T_2 spectrum of less than 16 ms is more than 95% of the total area of the T_2 spectrum in sample S. The T_2 spectrum of more than 16 ms occupies the rest 5% of the T_2 spectrum area because of the relatively large pore space in sample S. Thus, T_2 cutoff value is determined as 16 ms, given that the rest 5% in the total T_2 spectrum area is trivial. This means that T_2 spectrum at $T_2 < 16$ ms is contributed by water in the matrix while at $T_2 > 16$ ms is related to the water in the glass bead-filled fracture.

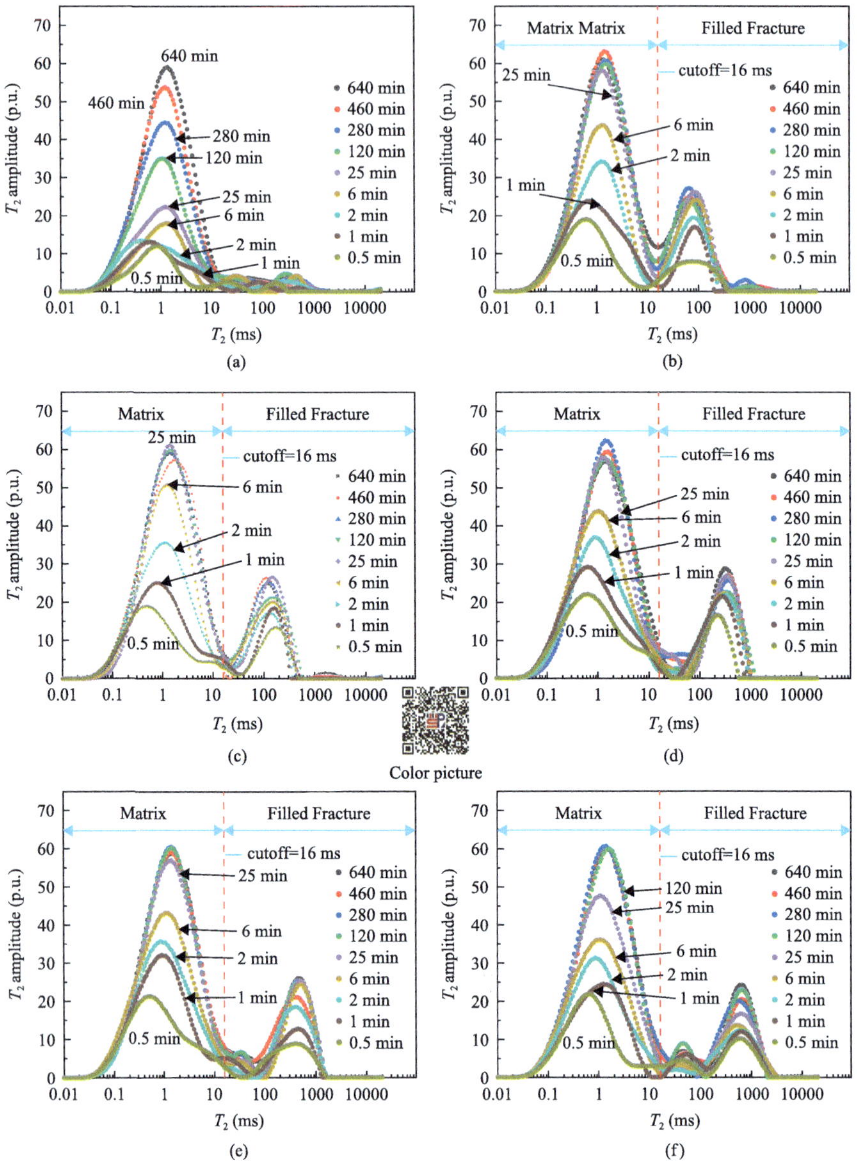

Fig. 7.2 The NMR T_2 spectrum obtained during imbibition in single- and dual-porosity media

Once the T_2 cutoff value is determined, it is possible to distinguish the dynamic distribution of the imbibed water from the matrix and the glass bead-filled fracture. Recalling the T_2 spectrum in Fig. 7.2, the water saturation of the matrix (e.g., the rate of matrix imbibition) in samples D_1–D_5 during the SI process can be easily determined.

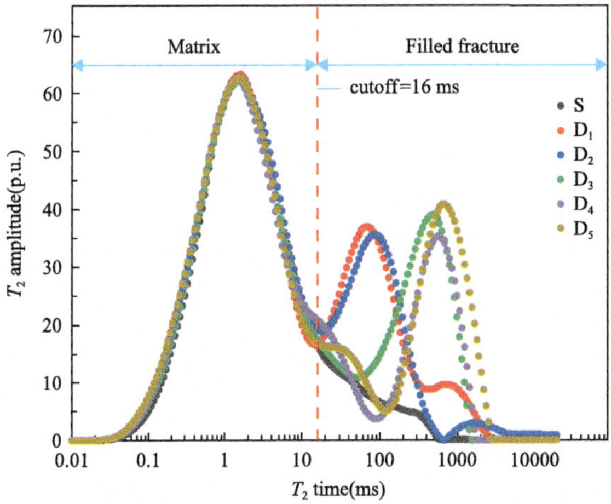

Color picture

Fig. 7.3 Comparison of the NMR T_2 spectrums under saturated condition

In Fig. 7.4, although the matrix is the same in samples S and D_1–D_5, the rate of matrix imbibition is inconsistent between the single-porosity media and the dual-porosity media. As is shown in Fig. 7.4, the rate of matrix imbibition in samples D_1–D_5 is much greater than that in sample S until the imbibed water saturates the matrix. In samples D_1–D_5, the imbibed water saturates the matrix after 120 min, while the matrix in sample S is saturated after 640 min. The accelerated rate of the matrix imbibition in samples D_1–D_5 can be attributed to the presence of the glass bead-filled fractures, since the filled fractures increase the contact area between the matrix and the water. Therefore, we can confirm that the matrix imbibition in the dual-porosity media is enhanced by the mass transfer between the matrix and the glass bead-filled fracture. Moreover, the water saturation of the matrix in samples D_1–D_5 is different at the same time, indicating that the mass transfer between the matrix and the glass bead-filled fracture depends on the particle sizes of the glass bead filled in fracture.

To quantitatively analyze the inconsistent rate of matrix imbibition and divide different imbibition stages in the SI process, there are two assumptions to simplify the problem. First, it is assumed that the water is imbibed into the matrix in a piston-like manner. Second, the pressure gradient in the gas phase is neglected. Therefore, the Handy model can be used to describe the imbibition process in the matrix based on the above assumptions,

$$Q_{\mathrm{w}}^2 = \left(\frac{2P_{\mathrm{c}}K_{\mathrm{w}}S_{\mathrm{w}}\phi A^2}{\mu_{\mathrm{w}}}\right)t \tag{7.3.1}$$

where Q_{w} is the volume of the imbibed water in the matrix, t is the imbibition time; μ_{w} is the viscosity of water, P_{c} is the capillary pressure at the wetting front, K_{w} is the effective hydraulic conductivity, ϕ is the porosity of the medium, and A is the

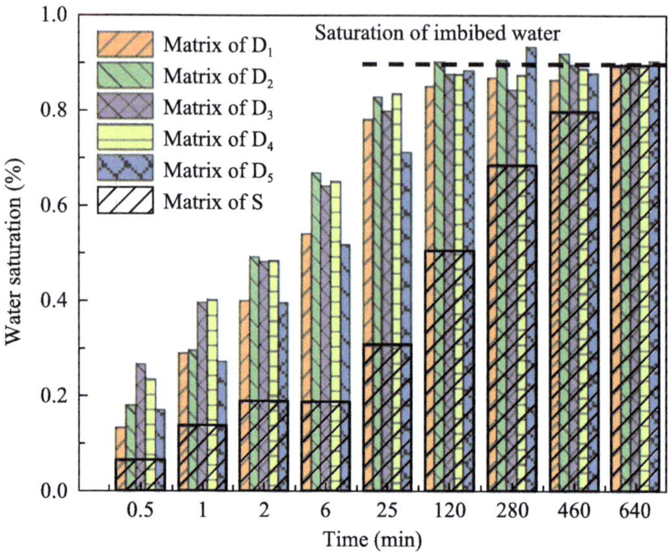

Fig. 7.4 The imbibition water saturation of the matrix in single-porosity (matrix only) and the dual-porosity media (both matrix and glass bead-filled fracture)

contact area between matrix and water. For sample S, the contact area A equals to the bottom area of the sandstone core A_0 (e.g., $A = A_0$), while for samples D_1–D_5, the contact area A equals to the sum of bottom area A_0 and the fracture contact area A_t (e.g., $A = A_0 + A_t$). Note that the fracture contact area A_t increases with time t at the beginning of the SI process. Furthermore, considering that the parameters μ_w, ϕ, P_c, S_w and K_w are constants for samples S and D_1–D_5, the Handy model can be reduced to:

$$Q_w = ADt^{1/2} = kt^{1/2} \tag{7.3.2}$$

where k is the ratio of matrix imbibition volume Q_w to square root of imbibition time $t^{1/2}$, which can be used as a measure of the imbibition rate, and D represents the absorptivity of the sandstone matrix and is a constant for samples S and D_1–D_5 which can be derived by

$$D = \sqrt{\frac{2P_c K_w S_w \phi}{\mu_w}} \tag{7.3.3}$$

According to the Handy model form Eq. (7.1.3), it can be inferred that the volume of the imbibed water in the matrix Q_w should be proportional to the square root of the imbibition time $t^{1/2}$. The measured relationship between Q_w and $t^{1/2}$ during the SI experiments is shown in Fig. 7.5. As is shown in Fig. 7.5a, the measured Q_w in the early-time stage of the SI process generally follows the proportional relationship with $t^{1/2}$, which is consistent with the Handy model from Eq. (7.1.3). In Fig. 7.5a, it is also found that the slope of the fitted line k for sample S is smaller than k for samples D_1–D_5, indicating that the rate of matrix imbibition is not only affected by the presence of the filled fractures but also sensitive to the particle sizes of the glass beads filled in the fracture. From the perspective of mass transfer between the matrix and the glass bead-filled fracture, the smaller k for sample S can be attributed to the smaller contact area $A(=A_0)$, while the larger k for samples D_1–D_5 is caused by larger contact area $A(=A_0 + A_t)$ where A_t was dynamically increased due to the rapid imbibition of the filled fracture during the SI process.

Based on the relationship between Q_w and $t^{1/2}$ throughout the SI process in Fig. 7.5b, the characteristics of the measured matrix imbibition can be divided into three stages as follows,

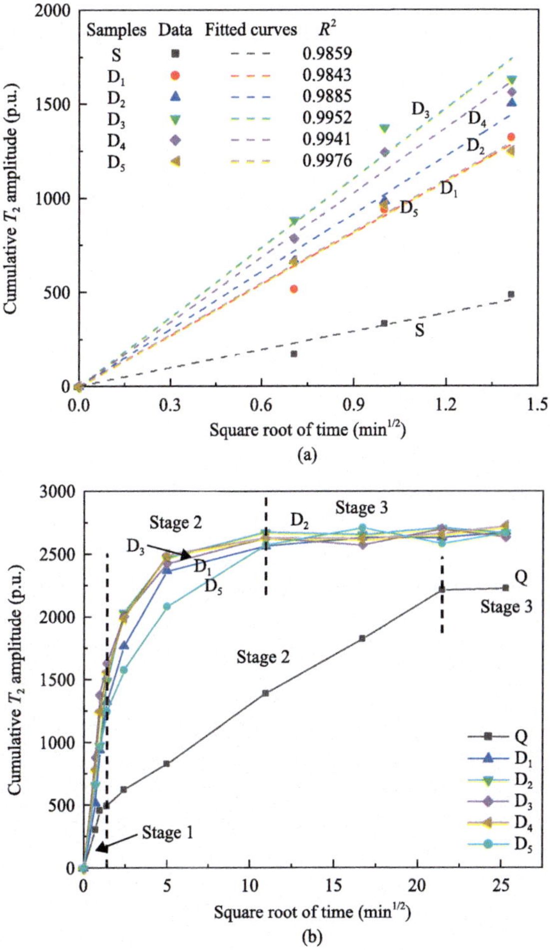

Fig. 7.5 Comparison of Q_w-$t^{1/2}$ relationship between single- and dual-porosity media, **a** at the early-time stage of the SI process and **b** during the entire SI process

(1) Early stage (Stage 1). In this stage, the influence of gravity can be neglected and the SI process can be characterized by the Handy model [Eq. (7.1.3)]. Thus, Q_w is linearly correlated with $t^{1/2}$ at this stage. And the imbibition rate of sample S is smaller than the imbibition rate of samples D_1–D_5 due to the smaller matrix-water contact area A.
(2) Middle stage (Stage 2). In this stage, the volume of the imbibed water continues to increase and the influence of gravity becomes more evident. Under the influence of gravity, the imbibition rate gradually decreases and consequently the slope of the Q_w-$t^{1/2}$ curve also decreases.
(3) Late stage (Stage 3). In this stage, the imbibition rate is close to zero and the volume of the imbibed water hardly increases as the medium gradually becomes saturated with the imbibed water.

7.4 Imbibition Rate and Analytical Model

Since the rate of matrix imbibition is not only affected by the presence of the filled fractures but also depends on the particle sizes of the glass bead filled in the fracture, a dimensionless imbibition rate ratio k_r is introduced to evaluate the influence of the particle sizes of the glass beads on the rate of matrix imbibition. The k_r is defined by

$$k_r = \frac{k_d}{k_s} \tag{7.4.1}$$

where k_d is the overall slope of Q_w-$t^{1/2}$ curve in the early stage of the SI in the dual-porosity media (samples D_1–D_5) and k_s is the overall slope of Q_w-$t^{1/2}$ curve in the early-time stage of the SI in the single-porosity medium (sample S).

The calculated k_r values at the early-time stage of the SI process are listed in Table 7.2. Table 7.2 shows that all calculated k_r values for samples D_1–D_5 are greater than 1.0, indicating that on the one hand, the presence of a filled fracture significantly enhances the SI process of dual-porosity medium. On the other hand, a clear non-monotonic relationship is found between k_r and the particle size of the glass beads.

As is shown in Table 7.2, as the particle size of the glass beads increases from 0.071–0.090 to 0.500–0.710 mm, k_r first increases from 2.11 to 2.82 and then

Table 7.2 The dimensionless imbibition rate ratio k_r at the early stage of imbibition

Medium type	Sample	k_r
Single-porosity medium	S	1.0
Dual-porosity medium	D_1	2.11
	D_2	2.49
	D_3	2.82
	D_4	2.79
	D_5	2.12

decreases to 2.12. This non-monotonic dependence of k_r on the particle size of the glass beads suggests that a potentially optimum particle size of the glass beads can most effectively enhance the rate of matrix imbibition which is originally limited due to low permeability and porosity.

The SI process is analyzed in terms of Eq. (7.1.3) for a dual-porosity porous medium for understanding the mechanism of the enhanced rate of matrix imbibition. As mentioned above, for the case of a dual-porosity medium, the contact area A in Eq. (7.1.3) consists of two components (i.e., the bottom area A_0 and the fracture contact area A_t). During the SI process, the bottom area A_0 is a constant and does not change with time, while the fracture contact area A_t increases with time and is closely correlated with the imbibition height in the fracture in the early-time stage of the SI process (see Fig. 7.3). Thus, considering the filled fracture as a bundle of capillary tubes, the SI process in the filled facture can be described by the LW law

$$h = \begin{cases} \sqrt{\dfrac{\lambda\sigma\cos\theta}{4\mu_w}}t^{\frac{1}{2}}, & t \le t_{max} \\ h_{max}, & t > t_{max} \end{cases} \tag{7.4.2}$$

where h is the imbibition height, t is the imbibition time, λ is the equivalent diameter of the capillary tubes, σ is the surface tension of water, θ is the contact angle for water and glass beads, h_{max} is the maximum height of the capillary imbibed water, and t_{max} is the time when the height of the capillary imbibed water reaches h_{max}. h_{max} and t_{max} can be calculated by

$$\begin{cases} h_{max} = \dfrac{4\sigma\cos\theta}{\lambda\rho g} \\ t_{max} = \dfrac{4\mu_w h_{max}^2}{\lambda\sigma\cos\theta} \end{cases} \tag{7.4.3}$$

According to previous studies (Czachor 2006; Li et al. 2015), when the particle size of the glass beads reaches 0.045–0.500 mm, the contact angle between water and glass beads stands at 75.2°–80.3°(e.g., $0.99985 < \cos\theta < 0.99987$). Thus, the value of $\cos\theta$ is set as 0.99986 in the following calculations.

Based on Eq. (7.4.2), the fracture contact area A_t can be determined by

$$A_t = \pi d_0 h = \begin{cases} \pi d_0\sqrt{\dfrac{\lambda\sigma\cos\theta}{4\mu_w}}t^{1/2}, & t \le t_{max} \\ \pi d_0 h_{max}, & t > t_{max} \end{cases} \tag{7.4.4}$$

where d_0 is the diameter of the cylindrical fracture. Considering that

$$M = \pi d_0 \sqrt{\frac{\lambda \sigma \cos\theta}{4\mu_w}} \tag{7.4.5}$$

is determined by the structure of the fracture (M increases with the fracture diameter d_0 and the equivalent capillary diameter λ) and does not change with time t, Eq. (7.4.4) can be simplified as follows,

$$A_t = \begin{cases} Mt^{1/2}, & t \le t_{max} \\ \pi d_0 h_{max}, & t > t_{max} \end{cases} \tag{7.4.6}$$

Substituting Eqs. (7.4.5) and (7.4.6) into LW law yields

$$Q_w = \begin{cases} DA_0 t^{1/2} + \frac{DMt}{2}, & t \le t_{max} \\ D(A_0 + \pi d_0 h_{max}) t^{1/2} - \frac{DMt_{max}}{2}, & t > t_{max} \end{cases} \tag{7.4.7}$$

The proposed Eq. (7.4.7) analytically indicates that if the matrix properties A_0 and D are considered as constants, the relationship between Q_w and $t^{1/2}$ in a dual-porosity porous medium is determined by the structure of the fractures. More especially, when $t \le t_{max}$, the relationship between Q_w and $t^{1/2}$ is controlled by M; when $t > t_{max}$, the relationship between Q_w and $t^{1/2}$ is determined by both h_{max} and t_{max}. All parameters M, h_{max} and t_{max} are controlled by the equivalent capillary diameter λ of the filled glass beads. Thus, the relationship between Q_w and $t^{1/2}$ in a dual-porosity medium can be predicted using Eq. (7.4.7).

The comparison of the experimentally measured and theoretically predicted Q_w-$t^{1/2}$ relationships is shown in Fig. 7.6, and the parameters used for the theoretical prediction are listed in Table 7.3. It should be mentioned that the h_{max} in Eq. (7.4.7) is reduced by a reduction factor of 0.54 to represent the influence of gravity on the maximum capillary imbibition height. From Fig. 7.6, it can be seen that the slope between Q_w and $t^{1/2}$ is not a constant even for a given dual-porosity porous media. This change of Q_w-$t^{1/2}$ relationship with time can be attributed to the variation in the fracture contact area A_t. When $t \le t_{max}$, A_t increases linearly with $t^{1/2}$, while $t > t_{max}$, A_t remains a constant (i.e., $A_t = \pi d_0 h_{max}$ when $t > t_{max}$). Since Q_w is linearly correlated with $A_t t^{1/2}$, it can be concluded that the relationship between Q_w and $t^{1/2}$ is a quadratic relationship when $t \le t_{max}$; while $t > t_{max}$, Q_w follows a linear function of $t^{1/2}$.

It can also be found from Fig. 7.6 and Table 7.3 that as the equivalent capillary diameter λ increases from 0.0181 to 0.1361 mm, the overall slope k_r between Q_w and $t^{1/2}$ initially increases and then decreases with λ. An increasing k_r between Q_w and $t^{1/2}$ indicates the increasingly promoting influence of the filled fracture on the matrix imbibition with increasing λ, while a decreasing k_r between Q_w and $t^{1/2}$ indicates decreasingly promoting influence of the filled fracture when λ becomes too large.

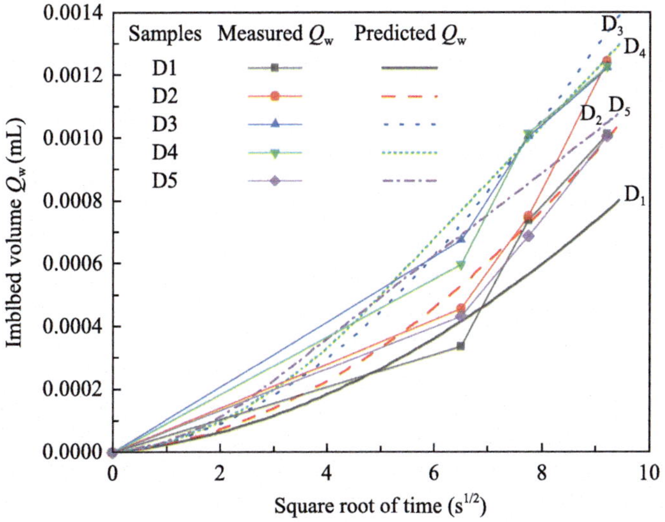

Fig. 7.6 Comparison of the experimentally measured and theoretically predicted Q_w-$t^{1/2}$ relationships

Table 7.3 The parameters used for predicting the Q_w-$t^{1/2}$ relationships in the experiments

Samples	Λ (mm)	d_0 (mm)	D [mL/(mm^2·min$^{1/2}$)]	μ_w (10^{-3} Pa·s)	σ (N/m)
D_1	0.0181	6.0000	0.0427	0.8937	0.0720
D_2	0.0343				
D_3	0.0737				
D_4	0.0962				
D_5	0.1361				

Thus, it can be inferred that there is an optimum λ that strongly enhances the matrix imbibition. For a given imbibition time T, the partial derivative of Q_w with respect to λ is

$$\frac{\partial Q_w}{\partial \lambda} = \begin{cases} \frac{\pi D d_0 T}{4}\sqrt{\frac{\sigma \cos\theta}{\mu_w}}\lambda^{\frac{1}{2}}, & \lambda \le \lambda_T \\ \frac{\pi D d_0 \sigma \cos\theta T^{\frac{1}{2}}}{\rho g}\lambda^{-\frac{7}{2}}(5\lambda_T^{\frac{3}{2}} - 4\lambda^{\frac{3}{2}}), & \lambda > \lambda_T \end{cases} \tag{7.4.8}$$

where

$$\lambda_T = 4\left(\frac{\sigma \mu_w \cos\theta}{\rho^2 g^2 T}\right)^{\frac{1}{3}} \tag{7.4.9}$$

λ_T is the equivalent capillary diameter under which the capillary imbibition height in the fracture reaches its maximum height h_{max} just at time T. From Eq (7.4.8), when $\lambda \leq \lambda_T$, $\partial Q_w/\partial\lambda$ is always greater than zero, which means that the imbibition volume increases with λ. And when $\lambda > \lambda_T$, $\partial Q_w/\partial\lambda$ is first larger and then smaller than zero, which means that Q_w first increases with λ and then decreases. Combining the two intervals in Eq. (7.4.8), the critical λ for the specified imbibition time T can be calculated as $(5/4)^{2/3}\ \lambda_T \approx 1.16\lambda_T$. This means that when $\lambda_T \leq 1.16\lambda_T$, the imbibition volume increases with λ, while $\lambda_T > 1.16\lambda_T$, the imbibition volume decreases with λ. When λ approaches the optimum value of $1.16\lambda_T$, the imbibition volume reaches its maximum value at time T, meaning that the filled fracture enhances matrix imbibition most significantly at the λ of $1.16\lambda_T$.

According to Eq. (7.4.9), λ_T in the early-time stage of the SI process (i.e., $T =$ 84 s) equals to 0.0799 mm, and thus the optimum equivalent capillary diameter λ_0, with the best promoting influence, should be 0.0926 mm. From Table 7.3, it can be seen that $\lambda_0 = 0.0926$ is between the λ of samples D_3 and D_4, which explains why samples D_3 and D_4 have a better promoting influence on matrix imbibition than the other samples. Chapter 8 Influence of Wettability on Interfacial Area for Immiscible Liquid Inv.

References

Akin S, Kovscek A (1999) Imbibition studies of low-permeability porous media, SPE western regional meeting. Society of Petroleum Engineers, Richardson

Alyafei N, Al-Menhali A, Blunt MJ (2016) Experimental and analytical investigation of spontaneous imbibition in water-wet carbonates. Transp Porous Media. https://doi.org/10.1007/S11242-016-0761-4

Bendel P (1990) Spin-echo attenuation by diffusion in nonuniform field gradients. J Magn Reson 86(3):509–515

Chen Y, Zhou Z, Wang J et al (2021) Quantification and division of unfrozen water content during the freezing process and the influence of soil properties by low-field nuclear magnetic resonance. J Hydrol 602:126719

Cheng CL, Perfect E, Donnelly B et al (2015) Rapid imbibition of water in fractures within unsaturated sedimentary rock. Adv Water Resour 77:82–89

Coates GR, Xiao L, Prammer MG (1999) NMR logging: principles and applications. Haliburton Energy Services Houston

Connolly PRJ, Vogt SJ, Iglauer S et al (2017) Capillary trapping quantification in sandstones using NMR relaxometry. Water Resour Res 53(9):7917–7932

Czachor H (2006) Modelling the effect of pore structure and wetting angles on capillary rise in soils having different wettabilities. J Hydrol 328(3–4):604–613

Das BM (2019) Advanced soil mechanics. CRC Press, Boca Raton

Hu Y, Patmonoaji A, Xu H et al (2021) Pore-scale investigation on nonaqueous phase liquid dissolution and mass transfer in 2D and 3D porous media. Int J Heat Mass Transf 169:120901

Li G, Chen X, Huang Y (2015) Contact angle determined by spontaneous imbibition in porous media: Experiment and theory. J Dispersion Sci Technol 36(6):772–777

Lyu C, Ning Z, Chen M et al (2019) Experimental study of boundary condition effects on spontaneous imbibition in tight sandstones. Fuel 235:374–383

Meng M, Ge H, Ji W et al (2015) Monitor the process of shale spontaneous imbibition in co-current and counter-current displacing gas by using low field nuclear magnetic resonance method. J Nat Gas Sci Eng 27:336–345
Wang X, Yang Z, Li H et al (2010) Experimental study on pore structure of low permeability core with NMR spectra. J Southwest Petrol Univ (science & Technology Edition) 32(2):69–72
Yuan X, Yao Y, Liu D et al (2019) Spontaneous imbibition in coal: experimental and model analysis. J Nat Gas Sci Eng 67:108–121
Zhao Y, Xue S, Han S et al (2017) Effects of microstructure on water imbibition in sandstones using X-ray computed tomography and neutron radiography. J Geophysic Res Solid Earth 122(7):4963–4981
Zhao Y, Wu Y, Han S (2019) Water sorptivity of unsaturated fractured sandstone-fractal modeling and neutron radiography experiment. Adv Water Resour 130(2019):172–183

Chapter 8
Influence of Wettability on Interfacial Area for Immiscible Liquid Invasion

This chapter aims to investigate how wettability influences α_{nw} during non-wetting phase invasion in a single fracture with self-affine rough surfaces. The successive random additions technique and the lattice Boltzmann method are employed to construct self-affine rough fracture wall surfaces and simulate the evolution of α_{nw}, respectively. Two model verifications are used to verify the entry pressure for non-wetting phase invasion and wetting phase drainage in a fracture. Subsequently, the model is applied to investigate the effect of wettability on flow patterns and the p_c-S_w-α_{nw} relationship for four different contact angles. Finally, the correspondence of the results with published α_{nw}-S_w correlations for porous media is examined.

8.1 Interfacial Area for Immiscible Liquid Invasion

For the mass transfer process of residual DNAPL, the interface area is a very key parameter, because the efficiency of transferring quality from DNAPL to aqueous phase liquid is directly proportional to the interface area. At the same time, mastering the evolution law of interface area and accurately calculating the interface area is an essential link for evaluating, predicting, and repairing the fractured medium area polluted by residual DNAPL. The early research on interface area model mainly focuses on porous media.

The biggest advantage of the interface area model based on thermodynamic equilibrium theory is that the interface area can be determined through the relationship between capillary pressure and saturation, However, this model has three shortcomings: ①It cannot calculate the effective interface area; ②The drainage history curve of the study area is not considered; ③The model assumes that all the work done by water pressure is transformed into surface free energy to form an aqueous liquid interface, without considering the influence of water pressure energy dissipation. By studying the change of interface area during the first drainage in porous media,

Z. Dou et al., *Mass Transfer Dynamics of Contaminants in Fractured Media*,
https://doi.org/10.1007/978-981-99-9187-7_8

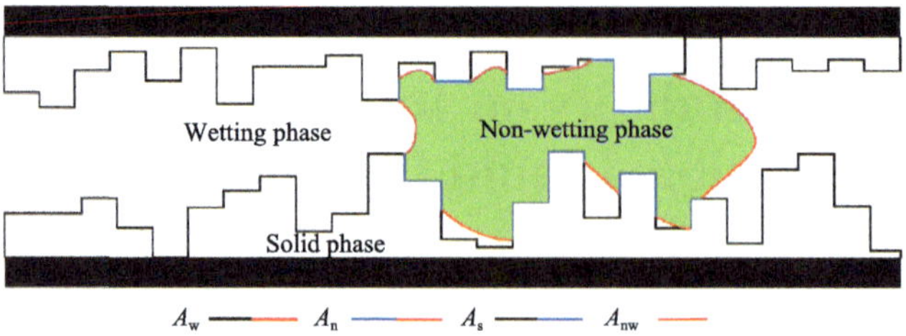

Color picture

Fig. 8.1 Illustration of interfacial areas in a rough fracture A_w includes the black and red lines, A_n includes the blue and red lines, A_s includes the black and blue lines, and the red line represents A_{nw}

through sensitivity analysis, it is found that hydraulic drawdown had an important impact on the intrusion form and residual distribution form of DNAPL. However, in addition to the distribution form of fillings, the difference of mineral composition of fillings will lead to the change of wettability. Although the above studies confirm that the characteristics of fracture medium structure are closely related to DNAPL mass transfer process, most are limited to a certain stage of DNAPL mass transfer process rather than the whole stage of DNAPL mass transfer process.

For a three-dimensional (3D) fracture, the specific interfacial area between non-wetting and wetting phases (also called effective interfacial area) is defined as $\alpha_{nw} = A_{nw}/v$, where v is the total fluid volume including both phases and A_{nw} is the total NAPL-water interfacial area in v. Similarly, the specific NAPL interfacial area is defined as $\alpha_n = A_n/v$ with A_n being the NAPL interfacial area (see Fig. 8.1), the specific water interfacial area is defined as $\alpha_w = A_w/v$ with A_w being the water interfacial area, and the specific solid surface area is defined as $\alpha_s = A_s/v$ with A_s being the solid surface area.

Many experimental approaches have been used to estimate α_{nw} including interfacial tracer techniques (Kim et al., 1997), X-ray computed tomography and microtomography (Wildenschild et al., 2002; Iglauer et al., 2012). However, conducting these methods is difficult, expensive, and time-consuming.

In addition, the theoretical models, such as thermodynamically-based models, involve estimation of interfacial areas based on an energy-balance approach and measured Pc-Sw curves for drainage (Leverett, 1941).

For imbibition, the thermodynamic model underestimates the interfacial area due to the assumption that entrapment occurs only within the largest pores (Dobson et al., 2006). However, this method is modified to allow the estimate of α_{nw} by incorporating consideration of saturation history, energy dissipation during drainage and imbibition (Bradford and Leij, 1997; Grant and Gerhard, 2007; Schroth et al., 2008).

8.2 Entry Pressure

A computational domain of 396 × 216 μm^2 representing a 2D variable aperture fracture is used for an initial verification test of the LBM implementation. The fracture is assumed to have contact points in the center that create three distinct apertures [$e_1 = 32.4$, $e_2 = 25.2$, and $e_3 = 14.4$ μm, respectively, see Fig. 8.2a]. Initially, the right section of the fracture is filled with non-wetting phase while the left section is occupied by the wetting phase. For the non-wetting phase invading the left section, the capillary pressure, serving as the difference between the non-wetting phase pressure and wetting phase pressure, has to exceed the displacement pressure associated with the largest aperture of the fracture, as given by:

$$p_{\text{th}} = \frac{2\gamma \cos\theta}{e} \tag{8.2.1}$$

where θ is the contact angle and e is the largest local aperture. In this case, the contact angle was set to $\theta = 45°$ and the influence of gravity was neglected. A constant velocity boundary condition is imposed on the right side of the computational domain, and a constant water pressure is imposed on the left boundary. In order to ensure that the capillary force dominates the two-phase flow, the capillary number is set to about 10^{-6} for all simulations.

Figure 8.2 shows the processes of the non-wetting phase invasion. Due to the constant inlet flux of NAPL on the right and constant water pressure boundary on the

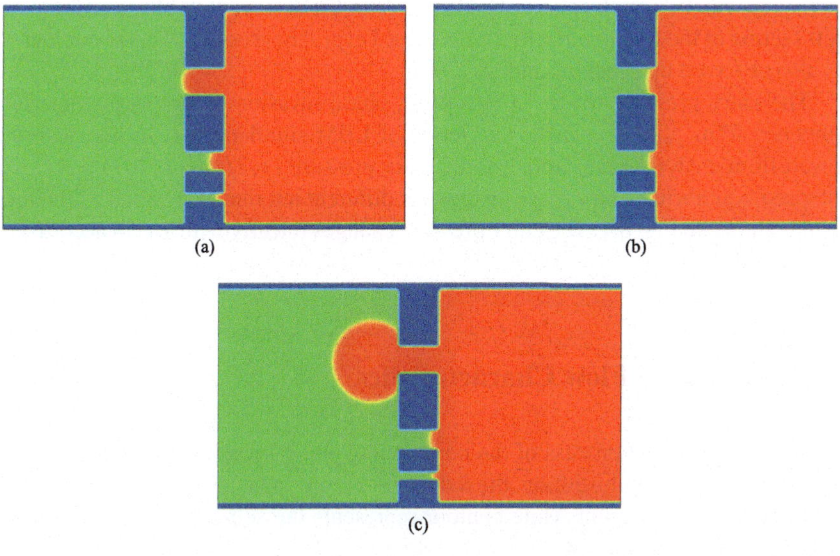

Fig. 8.2 Non-wetting phase invasion into the fractures with different apertures black domain represents solid phase, dark gray domain represents non-wetting phase, and light grey domain represents wetting phase **a** $t = 0.026$ s; **b** $t = 0.053$ s; **c** $t = 0.073$ s

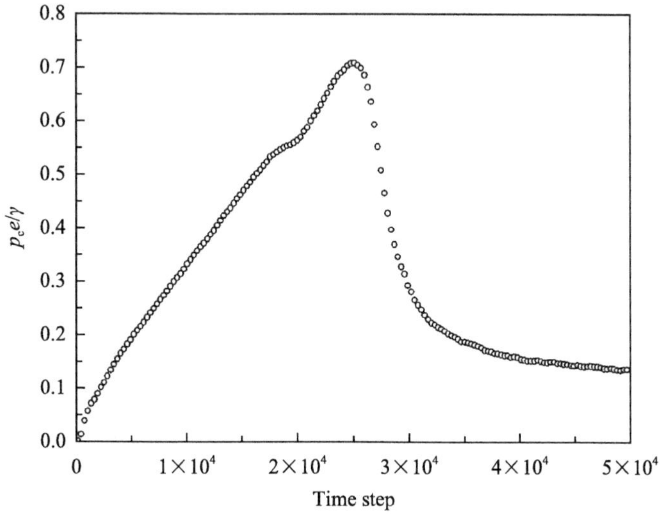

Fig. 8.3 Capillary pressure variation with time during the non-wetting phase invasion

left, capillary pressure increases with time. When the capillary pressure exceeds the displacement pressure for the largest aperture, NAPL moves through this aperture to the left side of the fracture.

This is accompanied by a sharp drop in the capillary pressure as is shown in Fig. 8.3. In the simulation, the interfaces between the non-wetting and wetting phases in the two smaller apertures recede slightly after the non-wetting phase flows to the left section of the fracture due to the decreased capillary pressure. This phenomenon is also observed by Moebius and Or (2012).

The entry pressure for the non-wetting phase invasion of the largest aperture obtained from $p_c e/\gamma = 0.7067$ in terms of LBM simulations is in line with the result from analytical non-dimensional entry pressure ($p_c e/\gamma = 0.7071$). A series of quasi-equilibrium entry pressure tests for different interfacial tensions with lattice unit is shown in Fig. 8.4. The entry pressure changes linearly with $1/e$, in accordance with Eq. (8.2.1).

8.3 Two Phase Flow Characteristics

Figure 8.5 shows the process of the non-wetting phase invading the fracture with $\theta = 45°$. For clarity of illustration, the red contour represents the invading non-wetting phase and the transparent white contour represents the self-affine surface, and the wetting phase is not shown.

The non-wetting phase follows two main paths corresponding to pathways with larger aperture, presenting overly qualitatively similar flow characteristics for all

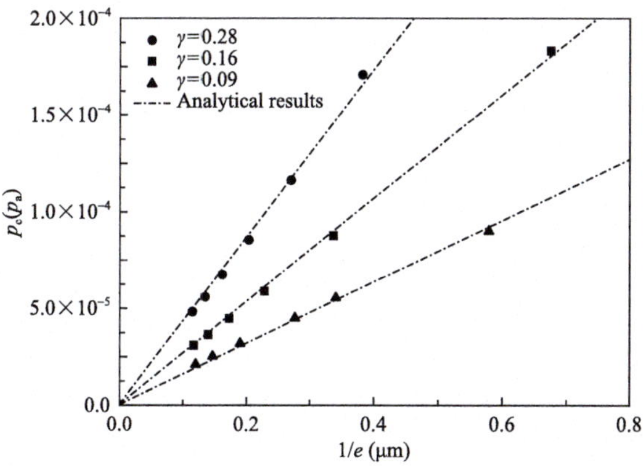

Fig. 8.4 Capillary pressure as a function of $1/e$ for different interfacial tensions

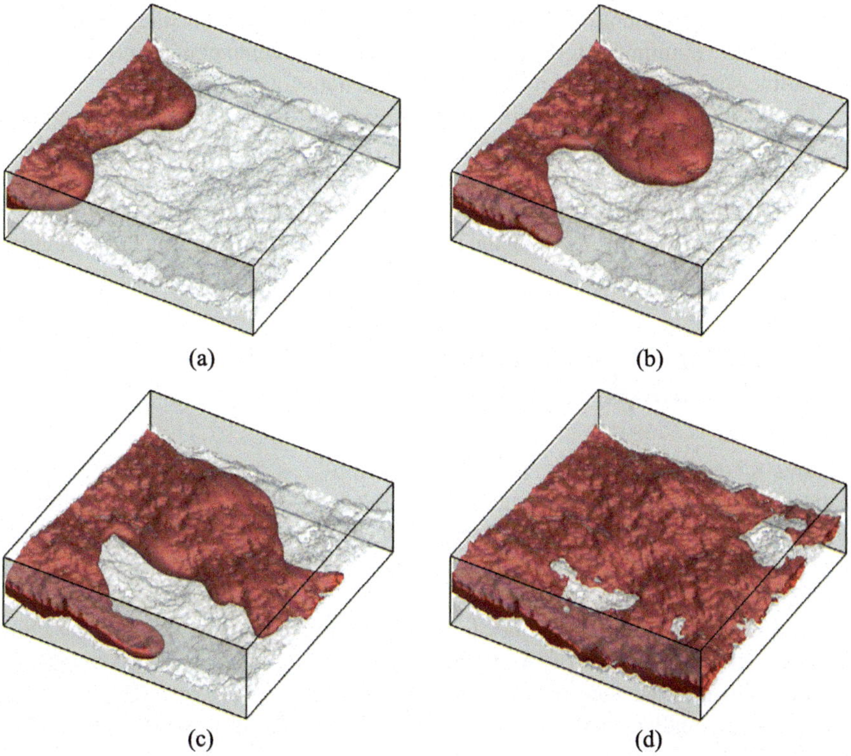

Fig. 8.5 Non-wetting phase invasion into the synthetic fracture for a contact angle of 45° **a** $t = 0.11$ s; **b** $t = 0.22$ s; **c** $t = 0.30$ s; **d** $t = 0.46$ s

contact angles. However, the contact angle does have an impact on the time to breakthrough and the average saturation at breakthrough. The non-wetting phase saturations at breakthrough time for four different contact angles of 65°, 45°, 25°, and 5° are 0.65, 0.61, 0.57 and 0.54, respectively. The durations to breakthrough for the four different contact angles of 65°, 45°, 25°, and 5° are 0.37, 0.30, 0.29, and 0.27 s.

As the contact angle decreases, the wettability of solid surface with respect to water increases. The increasing wettability leads to a smaller non-wetting phase saturation and a shorter time to breakthrough. The lower non-wetting phase saturation with increased wettability relies on the bypassing of a greater number of small aperture regions due to the increased entry pressure with lower contact angle.

In addition to the bypassing of small aperture regions, there is also substantially more water remaining as films on the upper and lower fracture surfaces as the contact angle is decreased, as is shown in Fig. 8.7.

Figure 8.7 shows the distributions of wetting phase at irreducible water saturation for the different contact angles. To clearly illustrate the trapped wetting phase distributions, the variable coloring of the water regions represents the elevation of a zone from the bottom fracture surface. The transparent white contour represents the self-affine surface, and the non-wetting phase is not shown.

The contact angle has a significant effect on the distributions of wetting phase at the end of drainage. It can be seen from Fig. 8.7 that there are two types of distribution of irreducible wetting phase in each case. One involves wetting phase trapped in isolated small aperture regions and the other involves irreducible wetting phase trapped as water films along the fracture wall surfaces in large aperture regions.

8.4 Capillary Pressure Saturation and Interfacial Area Relationships

Figures 8.7 and 8.8 show the changes in p_c and α_{nw} as functions of wetting phase saturation for different contact angles. When the non-wetting phase invades the fracture, non-zero α_{nw} and α_n are created and a_w decreases. As the capillary pressure increases and more non-wetting phase invades, the non-wetting phase follows the pathways that have large apertures (see Fig. 8.5).

As non-wetting phase continues to invade the fracture, some non-wetting phase breaks through the outlet of the fracture [see Figs. 8.4a and 8.5]. However, α_{nw} continues to increase (see Fig. 8.9) as the non-wetting phase invades the fracture regions with smaller apertures and new non-wetting phase interfaces form as the capillary pressure continues to increase.

As more pathways are invaded by the non-wetting phase, more NAPL-water interface forms. However, after reaching the peak value of α_{nw}, there is a rapid decrease as the irreducible water saturation approaches and the number of interfaces between wetting and non-wetting phases is decreased due to the displacement of all but the trapped water. This trend in the S_w-α_{nw} curve is qualitatively consistent with the experimental and numerical results for porous media reported by Porter et al. (2009) and Raeesi and Piri (2009).

Decreasing the contact angle increases the non-wetting phase entry pressures across the fracture, with the largest absolute changes in entry pressures occurring in the smallest aperture regions. This increase in entry pressures as contact angle is decreased, leading to an increase in capillary pressure (see Fig. 8.8) for a given water saturation. This trend is consistent with the experimental study (Drake et al., 2013). As non-wetting phase invades the fracture, regions surrounded by smaller aperture regions can become encircled by non-wetting phase and isolated from the bulk water phase. At this point the water in these regions is trapped and contributes to the irreducible water saturation.

In addition to water trapped in isolated small aperture regions, water remains as a film on the fracture walls and is trapped in depressions on the top and bottom fracture walls, as clearly evident in Figs. 8.6 and 8.7 that show the distribution of water in the fracture for each contact angle modeled. The presence of water as films in pits and crevices in rough fracture walls was discussed by Or and Tuller (2000). The lower contact angle leads to not only the higher entry pressure, but also stronger water–solid interaction strength that leads to thicker and more extensive water films.

The rate of increase of α_{nw} with decreasing water saturation is greater for lower contact angles. When the majority of fracture regions have been invaded by the non-wetting phase, the maximum α_{nw} values for the four different contact angles are equal to 0.562, 0.518, 0.454 and 0.356 μm^{-1}, and the corresponding S_{w} values are 0.26, 0.23, 0.22 and 0.18, respectively. Thus, the results show that the maximum α_{nw} increases as the contact angle decreases, while the water saturation at which the maximum α_{nw} occurs decreases as the contact angle increases. Furthermore, this result indicates that the entry pressure has a significant effect on the maximum α_{nw} during the primary drainage process, due to the fact that the contact angle decreases also results in the increment of entry pressure for a primary drainage process (Drake et al., 2013).

The maximum interfacial area occurs at saturations close to the irreducible water saturation where the capillary pressure increases sharply with decreasing water saturation. The sharply increasing capillary pressure causes the drainage of wetting phase in the pathways with relatively small apertures and further displacement of the wetting phase. As a result, the value of α_{nw} falls as the S_{w} approaches the irreducible water saturation. The non-zero value of the irreducible water saturation results directly in the non-zero value of α_{nw} at the end of the invasion of non-wetting phase, which is consistent with other published experimental and numerical studies (Bradford and Leij, 1997; Grant and Gerhard, 2007; Porter et al., 2009, 2010). For the four cases, the α_{nw} values at irreducible water saturation($S_{\text{wr}} = 0.04$, 0.06, 0.08 and 0.13)are equal to 0.035, 0.065, 0.102 and 0.132 μm^{-1} for $\theta = 65°$, 45°, 25° and 5°, respectively. This result also shows that the irreducible water saturation increases with decreasing contact angle, which is consistent with other published studies (Al-Futaisi and Patzek, 2003; Gharbi and Blunt, 2012).

Although it is typically assumed that non-wetting phase will not be trapped during drainage, Rossen (2003) showed how non-wetting phase could be trapped during

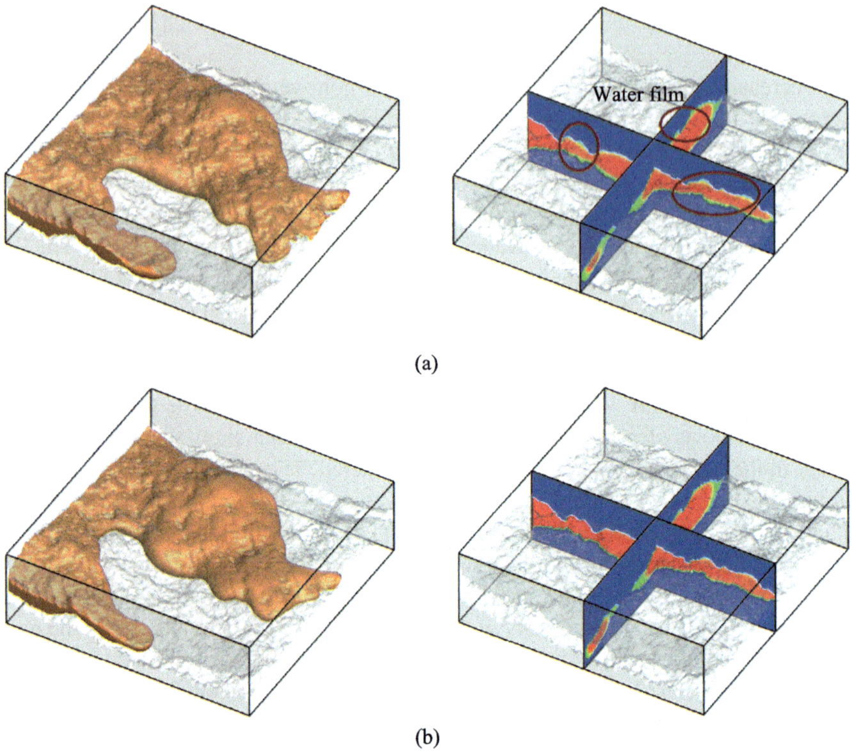

Fig. 8.6 Flow patterns and water films when non-wetting phase breakthrough occurs For the cross-section (the upper row), black represents fracture, gray represents the invading non-wetting phase, and light gray represents the wetting phase **a** $\theta = 5°$ and $t = 0.27$ s; **b** $\theta = 65°$ and $t = 0.37$ s

decreases in capillary pressure. In the LBM simulations the capillary pressure applied across the fracture is increasing, but the local capillary pressure is still fluctuating due to the movement of the non-wetting phase through the non-uniform distribution of apertures. This can lead to backwards flow of water in filaments (Rossen, 2003) and entrapment of non-wetting phase. This is not a significant phenomenon in the simulations, but a few regions where non-wetting phase is completely surrounded by wetting phase are observed in the simulation results, in the white areas in Fig. 8.5d that correspond to large areas otherwise occupied by the wetting phase.

The thermodynamically-base interfacial area model presented by Grant and Gerhard(2007) (which they refer to as the explicit interfacial area sub-model) for primary drainage in porous media is employed to fit the α_{nw} and S_{w} curves for the different contact angles:

$$\alpha_{\mathrm{nw}}(S_{\mathrm{w}}) = \psi(S_{\mathrm{w}}) \cdot E_{\mathrm{d}} \cdot \varphi \cdot \frac{\Phi_{\mathrm{nw}}(S_{\mathrm{w}})}{\gamma} \tag{8.4.1}$$

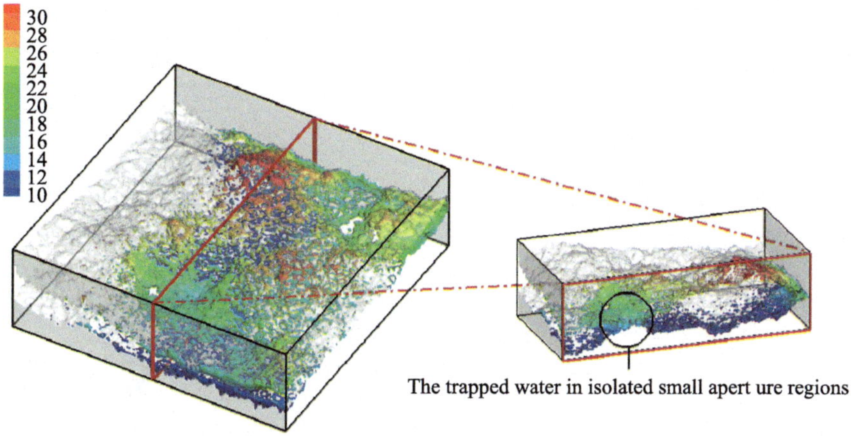

(a)

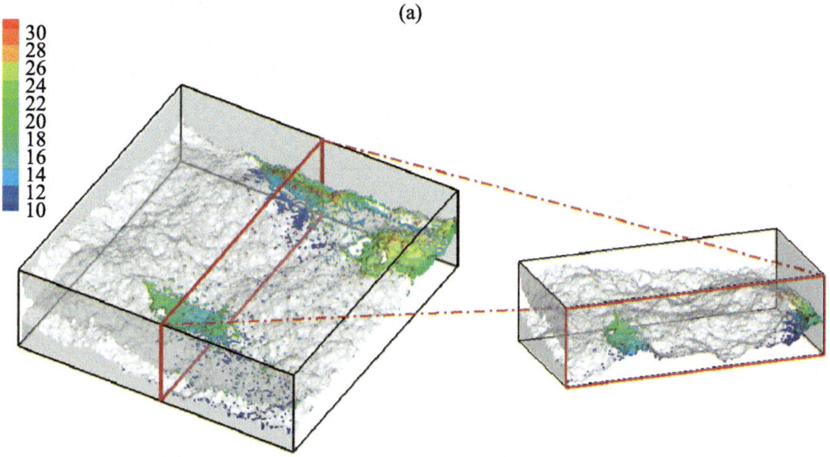

(b)

Fig. 8.7 Distributions of irreducible wetting phase for the contact angles of 5° and 65°. The variable color contour legend represents distance from the bottom fracture surface, expressed as lattice units **a** $\theta = 5°$; **b** $\theta = 65°$

where $\psi(S_w)$ is the empirical ratio between the α_{nw} and α_n, E_d is the energy dissipation factor, $\Phi_{nw}(S_w)$ is the integral of the of p_c-S_w curves, and φ is the porous medium porosity. For a single rough fracture, φ is assumed to be one. Grant and Gerhard (2007) showed a fourth-order polynomial best fit function for the $\psi(S_w)$ in porous media. p_c-S_w curves shown in Fig. 8.8 are integrated to determine $\Phi_{nw}(S_w)$ for each contact angle. A fourth-order polynomial for $\psi(S_w)$ is determined that provides the best fit (see Fig. 8.10) to the α_{nw}/α_n values from the LBM simulations for the four different contact angles:

$$\psi(S_w) = 4.8161S_w^4 - 2.7742S_w^3 - 4.4176S_w^2 + 3.6527S_w - 0.0116 \qquad (8.4.2)$$

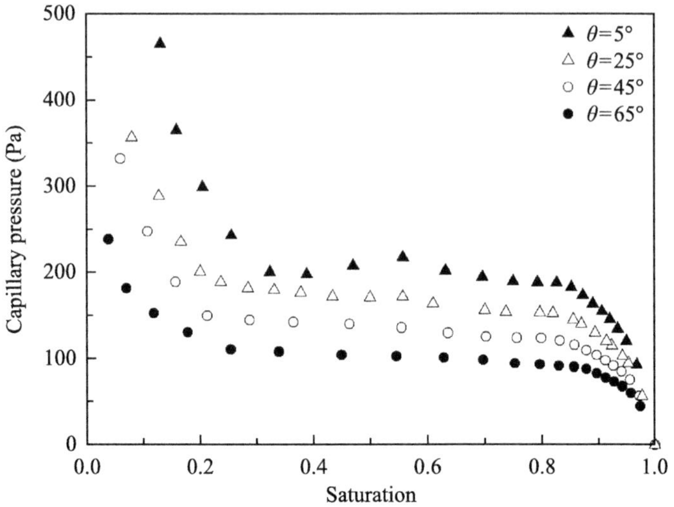

Fig. 8.8 Capillary pressure and saturation curves for different contact angles

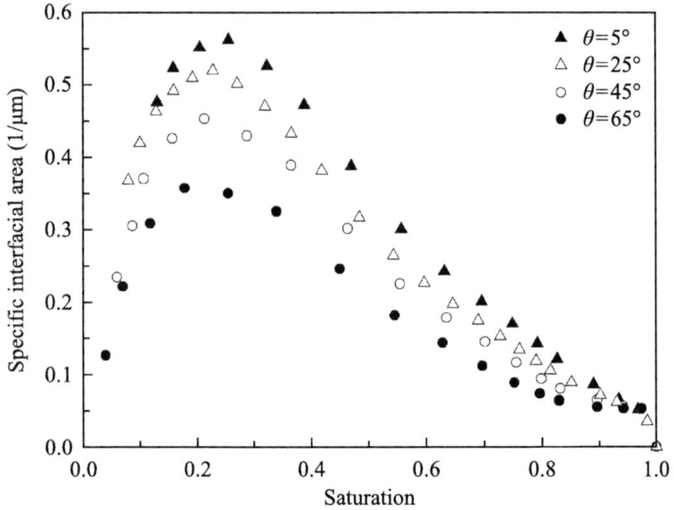

Fig. 8.9 Fluid–fluid interfacial area and saturation curves for different contact angles

Figure 8.10 shows that the $\psi(S_w)$ values for all 4 contact angles fall on the same curve and are well fit by Eq. 8.4.2. Figure 8.11 shows the LB and thermodynamic model results for the four different contact angles.

The thermodynamic model results shown in Fig. 8.11 are determined from Eq. (8.4.1) with integration of the p_c-S_w curves from the LBM predictions (Fig. 8.8), Eq. (8.4.2), and E_d values adjusted to get the best fit to the α_{nw}-S_w curves from the

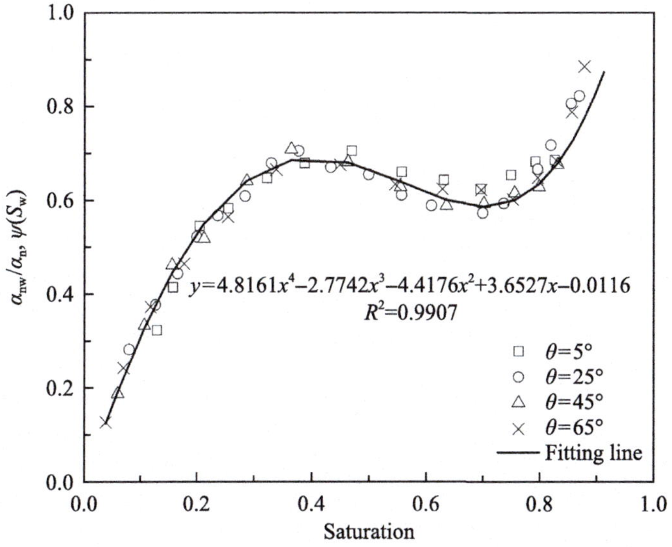

Fig. 8.10 Dependence of α_{nw}/α_n on saturation from LBM predictions and fourth-order polynomial fit

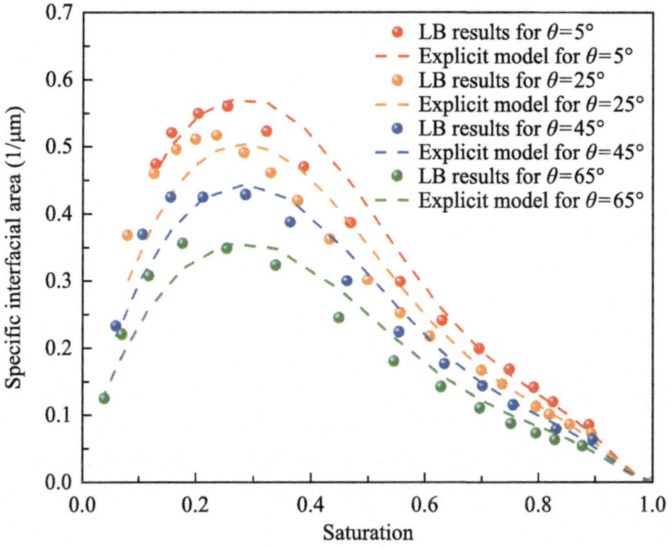

Fig. 8.11 LB and thermodynamic model α_{nw}-S_w results for the four different contact angles

LBM results. In general, the trends of the thermodynamic model results are consistent with the LBM results for different contact angles. In order to obtain the best-fit α_{nw} between the LBM results and thermodynamic model, the energy dissipation factors were set to $E_d = 0.86$, $E_d = 0.82$, $E_d = 0.75$, and $E_d = 0.7$ for $\theta = 65°$, $45°$, $25°$ and $5°$, respectively.

The energy dissipation factor is introduced into the explicit model to account for Haines Jumps. The energy dissipation for Haines Jumps results from the mechanical work done on the system not being completely converted into surface energy. For unconsolidated porous media, the energy dissipation factor is suggested to be $0.54 < E_d < 0.7$ by Dalla et al. (2002). Grant and Gerhard(2007)applied the explicit model to fit the experimental α_{nw} data set of Brusseau et al.(2010)and found that a single and constant value of $E_d = 0.21$ is sufficient to describe energy dissipation for the unconsolidated porous medium investigated. A smaller E_d is associated with the more energy losses in the system. Thus, $E_d = 0.7$ (the minimum value for the four different contact angles) for a single rough fracture suggests that smaller energy dissipations may occur in rough fractures than unconsolidated porous media from experiments and pore network models. In addition, Fig. 8.11 implies that the value of E_d is dependent on the contact angle for the cases examined.

Increased water film thickness also contributes to a decrease in the effective porosity for non-wetting phase invasion at any particular capillary pressure. The decreasing contact angle leads to the increased capillary pressure and increased irreducible water saturation due to greater retention of water in isolated regions, as well as films in pits and depressions on the top and bottom fracture surfaces.

The increased capillary pressure can be attributed to the direct influence of contact angle on capillary pressure. Decreasing contact angle also increases α_{nw} for a given water saturation. This is attributed to greater retention of water in low aperture regions as well as retention as water films, creating greater contact between water and non-wetting phase than will be created with more complete displacement of water by the invading non-wetting phase.

Fitting the thermodynamic model of Grant and Gerhard (2007) to the LBM model predictions indicates that the dependence of the ratio α_{nw}/α_n on S_w is not significantly affected by the contact angle and is well fit by a fourth-order polynomial, as Grant and Gerhard (2007) did for porous medium. The fit of the Grant-Gerhard model also shows that less energy dissipation will be expected to occur in two-phase flow rough fractures than has been reported for real porous media and pore network models. The best fit energy dissipation factors are dependent on contact angles in the simulated fracture.

References

Al-Futaisi A, Patzek TW (2003) Impact of wettability alteration on two-phase flow characteristics of sandstones: a quasi-static description. Water Resour Res 39(2):1042

Bradford SA, Leij FJ (1997) Estimating interfacial areas for multi-fluid soil systems. J Contam Hydrol 27(1–2):83–105

Brusseau ML, Narter M, Janousek H (2010) Interfacial partitioning tracer test measurements of organic-liquid/water interfacial areas: application to soils and the influence of surface roughness. Environ Sci Technol 44(19):7596–7600

Dalla E, Hilpert M, Miller CT (2002) Computation of the interfacial area for two-fluid porous medium systems. J Contam Hydrol 56(1–2):25–48

Dobson R, Schroth MH, Oostrom M et al (2006) Determination of NAPL-water interfacial areas in well-characterized porous media. Environ Sci Technol 40(3):815–822

Drake SS, O'Carroll DM, Gerhard JI (2013) Wettability contrasts between fresh and weathered diesel fuels. J Contam Hydrol 144(1):46–57

Gharbi O, Blunt MJ (2012) The impact of wettability and connectivity on relative permeability in carbonates: a pore network modeling analysis. Water Resour Res 48(12):W12513

Grant GP, Gerhard JI (2007) Simulating the dissolution of a complex dense nonaqueous phase liquid source zone: 1. model predict interfacial area. Water Resour Res 43(12):W12410

Iglauer S, Fernø MA, Shearing P et al (2012) Comparison of residual oil cluster size distribution, morphology and saturation in oil-wet and water-wet sandstone. J Colloid Interface Sci 375(1):187–192

Kim H, Rao PSC, Annable MD (1997) Determination of effective air-water interfacial area in partially saturated porous media using surfactant adsorption. Water Resour Res 33(12):2705–2711

Leverett MC (1941) Capillary behavior in porous solids. Transact AIME 142:152–169

Moebius F, Or D (2012) Interfacial jumps and pressure bursts during fluid displacement in interacting irregular capillaries. J Colloid Interface Sci 377(1):406–415

Or D, Tuller M (2000) Flow in unsaturated fractured porous media: Hydraulic conductivity of rough surfaces. Water Resour Res 3(5):1165–1177

Porter ML, Schaap MG, Wildenschild D (2009) Lattice-Boltzmann simulations of the capillary pressure–saturation–interfacial area relationship for porous media. Adv Water Resour 32(11):1632–1640

Porter ML, Wildenschild D, Grant G et al (2010) Measurement and prediction of the relationship between capillary pressure, saturation, and interfacial area in a NAPL-water-glass bead system. Water Resour Res 46(8):W08512

Raeesi B, Piri M (2009) The effects of wettability and trapping on relationships between interfacial area, capillary pressure and saturation in porous media: a pore-scale network modeling approach. J Hydrol 376(3–4):337–352

Rossen WR (2003) A critical review of Roof snap-off as a mechanism of steady-state foam generation in homogeneous porous media. Colloid Surface A 225(1–3):1–24

Schroth MH, Oostrom M, Dobson R et al (2008) Thermodynamic model for fluid–fluid interfacial areas in porous media for arbitrary drainage–imbibition sequences. Vadose Zone J 7(3):966

Wildenschild D, Vaz CMP, Rivers ML et al (2002) Using x-ray computed tomography in hydrology: systems, resolutions, and limitations. J Hydrol 267(3–4):285–297

Chapter 9
Multiscale Roughness Influence on Solute Transport in Fracture

This chapter aims: ①To investigate the influence of multi-scale roughness on linear, power-law, and second-order polynomial relationships between the D_L and *Pe*; ②To investigate the validation of the different relationships between the D_L and *Pe*. To do this, we generate a series of 2D variable-aperture fractures with the different Hurst exponents. A wavelet analysis method is implemented to decompose the roughness of the fracture wall into the primary roughness and secondary roughness. The BTCs numerically generated from direct simulations are fitted by two inverse models.

9.1 Statistical Self-affine Property

Many efforts have been made to characterize the morphology of natural fracture wall surfaces and indicate that the surface of natural fractures typically follows a statistical self-affine distribution (Mandelbrot 1983; Odling 1994). For the sake of completeness, we briefly review the mathematical characterization of the self-affine rough fracture wall. A more detailed generation method for the self-affine rough fracture wall can be found in our previous work (Dou et al. 2013). A 2D self-affine rough fracture wall height, $Z(x)$, satisfies

$$\lambda^H Z(x) = Z(\lambda x) \tag{9.1.1}$$

where H indicates the magnitude of the roughness or the so-called Hurst exponent varying from 0 to 1 and λ is a scaling factor. $Z(x)$ can be thought of as a function of an independent spatial or temporal variable x. For a self-affine rough fracture wall, the stationary increment $[Z(x+h_x) - Z(x)]$ over the distance h_x follows a Gaussian distribution with zero mean. Thus, for the arbitrary λ, the mean and variance of the increments can be expressed as

Z. Dou et al., *Mass Transfer Dynamics of Contaminants in Fractured Media*,
https://doi.org/10.1007/978-981-99-9187-7_9

$$\langle Z(x + \lambda h_x) - Z(x) \rangle = 0 \tag{9.1.2}$$

$$\sigma^2(\lambda) = \lambda^{2H} \sigma^2(1) \tag{9.1.3}$$

where $\langle x \rangle$ represents the mathematical expectation. For the different distances, the variance $\sigma^2(\lambda)$ is defined as a function of λ

$$\sigma^2(\lambda) = \langle [Z(x + \lambda) - Z(x)]^2 \rangle \tag{9.1.4}$$

$$\sigma^2(\lambda h_x) = \langle [Z(x + \lambda h_x) - Z(x)]^2 \rangle \tag{9.1.5}$$

Then, depending on Eq. (9.1.3), we have

$$\sigma^2_{\lambda h_x} = \lambda^{2H} \sigma^2_{h_x} \tag{9.1.6}$$

$$\sigma_{\lambda h_x} = \lambda^H \sigma_{h_x} \tag{9.1.7}$$

where $\sigma^2_{\lambda h_x}$ and $\sigma_{\lambda h_x}$ represent the variances of increments with different distances λh_x and λ, respectively.

The rough surface morphology of dolomite rock can be measured by a 3D stereo-topometric measurement system (Advanced Topometric Sensor, ATOS, manufactured by GOM mbH). A raw data set of fracture surface heights from a dolomite rock sample 280 × 210 × 70 mm is obtained by ATOS with a resolution of 10^{-3} μm. Over 1.6 million height points are captured in the raw data set. It should be noted that for height points, the increments of horizontal coordinates in x–y plane are non-grid and not constant. In order to analyze the scaling properties of dolomite rock surfaces, the raw data set is interpolated into about 5.6 million height points with a grid resolution of 100 μm.

Depending on the variogram analysis (Develi and Babadagli 1998), we evaluate the self-affinity of dolomite rock surface and examined the distribution of Hurst exponent values over 200 profiles along the longitude direction [for example, A-B profile in Fig. 9.1a]. The calculated Hurst exponent values distribute between 0.55 and 0.91. Figure 9.1b shows that the calculated Hurst exponent follows a Gaussian distribution with the mean $\mu_H = 0.70477$ and standard deviation $\sigma_H = 0.06832$.

As is known, for a Gaussian distribution, there is about 95% of the values within two standard deviations away from the mean ($\mu_H \pm 2\sigma_H$), which means that only 5% of the total calculated Hurst exponent is less than $\mu_H - 2\sigma_H = 0.56813$ or more than $\mu_H + 2\sigma_H = 0.84141$. Therefore, for saving the computational cost, we focus on the 2D self-affine fracture walls with $H = 0.6$, 0.7, and 0.8.

For 2D self-affine walls, the simplest method to generate a variable-aperture fracture is that the lower self-affine wall translates a distance e normal to the longitude direction. However, the corresponding local apertures in the rough fracture are constant and equal to e. Many experimental and numerical studies (Zimmerman et al.

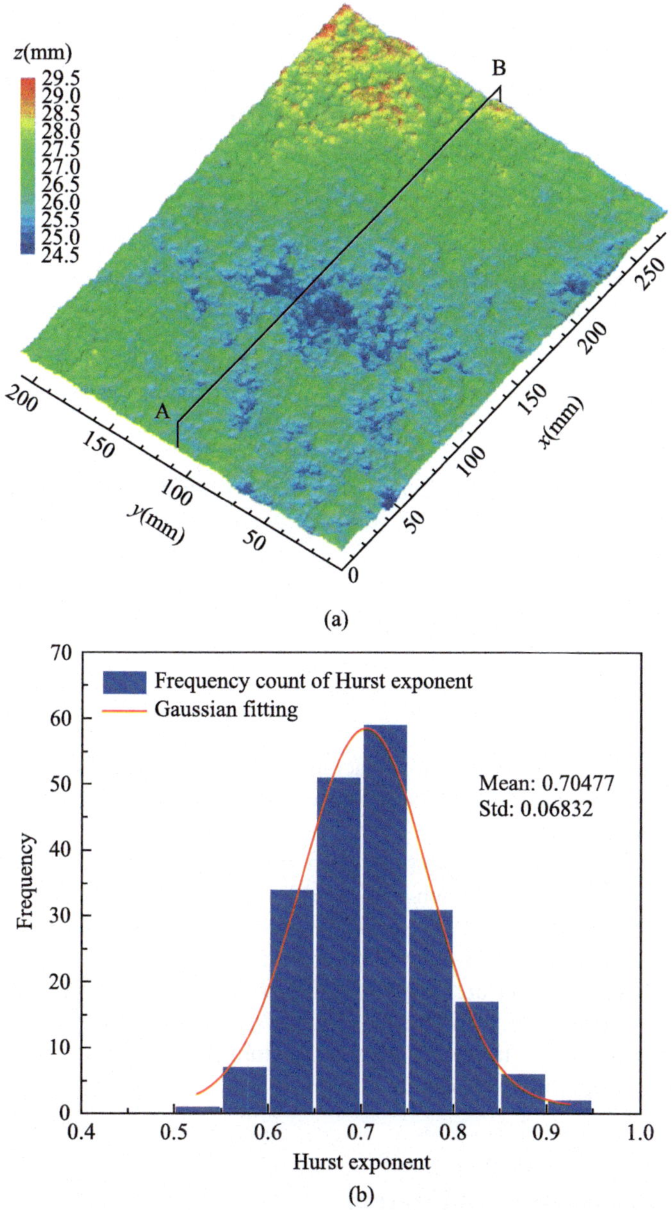

Fig. 9.1 Dolomite rock fracture surface 3D profile and distribution of Hurst exponent for different 2D profile along the longitude direction **a** Scanned rough surface morphology of dolomite rock; **b** Gaussian fit of the Hurst exponent of 2D profiles in the dolomite rock

1991; Wang and Cardenas 2014; Wang et al. 2015) have reported that the property of the fluid flow and compound transport in a fracture is highly sensitive to the variable aperture. Thus, alternatively, the shear displacement model proposed by Wang et al. (1988) is employed to obtain the local apertures with Gaussian distribution based on a self-affine wall. Assuming a 2D self-affine wall $S_1(x)$ with a position function,

$$S_2(x, y) = S_1(x + x_d) + \overline{b} \tag{9.1.8}$$

where x_d is the shear displacement and $\overline{b}$ is the mean aperture between top and bottom walls. The aperture as a function of top and bottom walls is defined as,

$$b(x) = \begin{cases} S_2(x) - S_1(x), & \text{if } S_2(x) > S_1(x) \\ 0, & \text{otherwise} \end{cases} \tag{9.1.9}$$

When $x_d = 0$ and $\overline{b}>0$, the fracture wall $S_1(x)$ is translated normal to the longitudinal direction with a distance $\overline{b}$, in which two rough walls can overlap and the corresponding aperture is constant $\overline{b}$. For $x_d \neq 0$ and $\overline{b}>0$, two rough walls of the fracture do not overlap and the aperture distribution obeys a Gaussian distribution with a mean of $\overline{b}$ and a standard deviation of σ,

$$f(b) = \frac{1}{\sqrt{2\pi\sigma}} e^{-\frac{(b-\overline{b})^2}{2\sigma^2}} \tag{9.1.10}$$

As shown in Fig. 9.2, the bottom fracture wall is a self-affine wall with $H = 0.6$ and the top fracture wall is generated by the shear displacement model with $x_d = 3$ mm and $\overline{b} = 0.5$ mm.

Figure 9.2 shows that the aperture distribution in the self-affine rough fracture follows the Gaussian distribution with $\sigma = 0.2$ mm and $\overline{b} = 0.5$ mm. It should be noted that the minimum aperture could be zero in 2D variable-aperture fractures when the fracture is generated by the shear displacement model.

In practice, zero-aperture region results in dead fractures or contact areas. However, we focused on the influence of multi-scale roughness on the relationship between the D_L and *Pe* in the 2D variable-aperture fractures. Thus, those fractures including zero-aperture were excluded.

9.2 Roughness Decomposition

The variable-aperture fracture walls can be thought of as signal profiles due to the local changes of geometries at different scales. Here, $f\ (\tau)$ is defined as the fracture wall profile. Analogy to the signal profiles, the f representing the variation of the wall height is similar with the variable in the frequency domain, while the τ representing

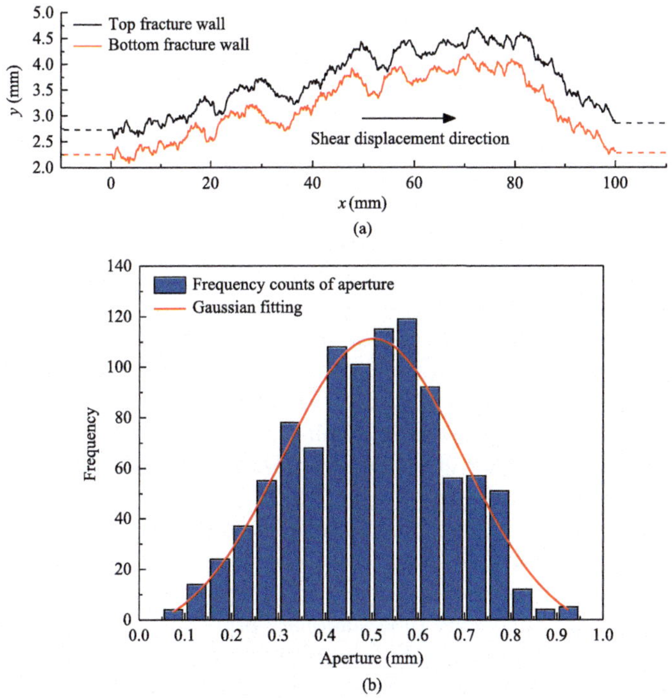

Fig. 9.2 A 2D Variable-apterture fracture with Hurst exponent of 0.6 **a** Illustration of the generation of 2D variable-aperture fracture with Hurst exponent of 0.6 by the shear displacement model; **b** Gaussian fitting of the aperture distribution in the generated 2D variable-aperture fracture with Hurst exponent of 0.6

the variation of the distance along in-plane direction of fracture is similar to the variable in the time domain. Thus, the roughness components can be distinguished and filtered with varying frequencies by mathematical methods, such as image analysis, Fourier transform, and wavelet analysis.

The wavelet analysis method has a natural advantage for analyzing and decomposing the roughness of the self-affine fracture wall by extracting the different frequencies of self-affinity geometric information at different frequency scales. The decomposition results obtained by using the wavelet analysis show that a self-affine variable-aperture fracture wall consists of non-stationary primary and stationary secondary roughness components (Zou et al. 2015).

The non-stationary primary roughness represents the lower frequency but larger scale undulations or waviness, while the latter one represents the higher frequency. The wavelet transform of the self-affine fracture profile can be expressed as,

$$W(a,l) = \int_{-\infty}^{+\infty} f(\tau)\psi_{a,l}(\tau)d\tau \tag{9.2.1}$$

where a is the scaling parameter representing the frequency of original fracture wall height profile, l represents the horizontal distance, and the function $\psi_{a,l}(\tau)$ is the so-called mother wavelet function expressed as,

$$\psi_{a,l}(\tau) = a^{-\frac{1}{2}}\psi\left(\frac{\tau - l}{a}\right) \tag{9.2.2}$$

An appropriate mother wavelet function is significant to decompose the roughness component of the self-affine fracture profile. The Daubechies' Wavelet (db8) is chosen as the mother wavelet. The process of roughness decomposition is realized by Mallat's pyramidal algorithm in a level-by-level procedure (Mallat 1989).

In every level procedure, each self-affine fracture wall is decomposed into low and high-frequency components. The low-frequency component is defined as an approximation wall and represents the dominant geometrical formation of waviness. The high-frequency component is defined as a detail wall and represents the small-scale waviness or unevenness.

As is shown in Fig. 9.3, the original rough fracture walls were decomposed and reconstructed by eight levels of decomposition profiles, noted by A_1–A_8 and D_1–D_8, respectively. It can be seen that at each level, the approximation walls, A_i, represent the primary waviness geometry (non-stationary primary roughness) and the detail walls, D_i, represent the small-scale unevenness (stationary secondary roughness). Thus, the original roughness can be decomposed into two different independent roughness, i.e. primary roughness and secondary roughness.

During the process of roughness decomposition, it is crucial to determine an appropriate cut-off level for separating roughness into the non-stationary primary roughness and the secondary roughness, where the non-stationary primary roughness should be dominant for the geometric property, while the secondary roughness can be treated as a Gaussian white noise process. Here, we use the quantitative criterion proposed by Zou et al. (2015). In Zou's quantitative criterion, the cut-off level is determined by analyzing the variance of the primary roughness, where the variance of the primary roughness becomes a remarkable change, while the secondary roughness can be considered as a Gaussian white noise process with the mean value closes to 0.

In Fig. 9.3, the cut-off levels of the top and bottom fracture walls are Level 3 and Level 4, respectively. Consequently, the Level 3 and 4 approximation walls are considered as the primary roughness for the top and bottom fracture walls, respectively. The secondary roughness of top and bottom fracture walls is determined by summing the Level 1–3 and Level 1–4 detail walls, respectively.

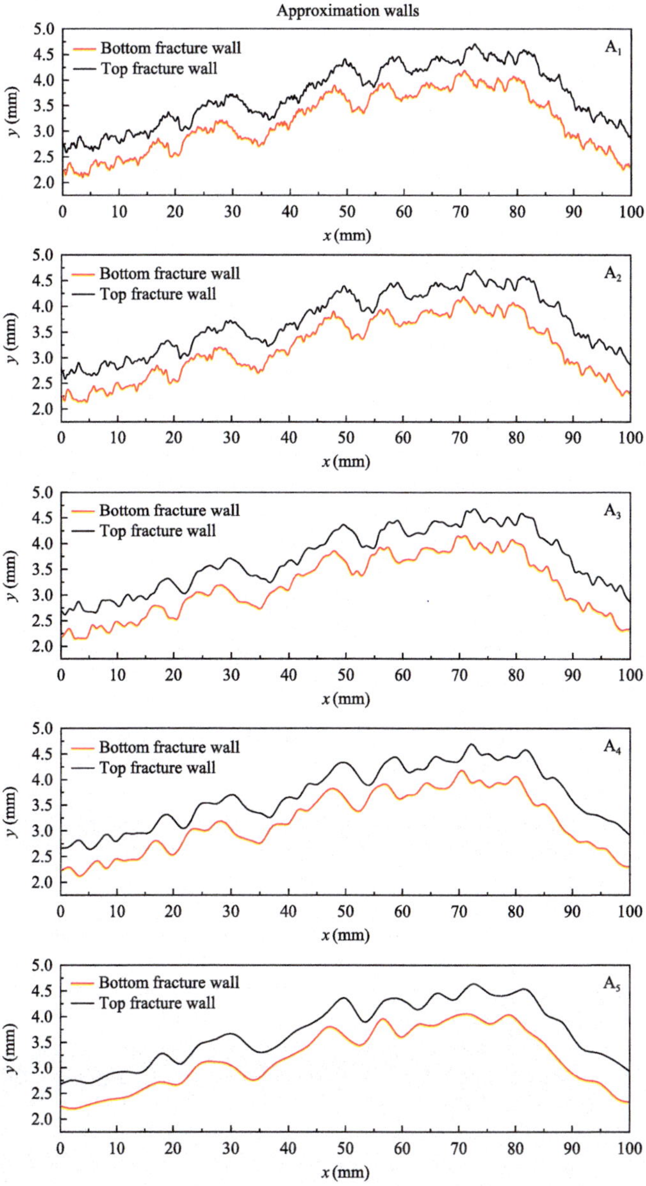

Fig. 9.3 Eight levels wavelet decomposition of the roughness for the self-affine fracture with $H = 0.6$

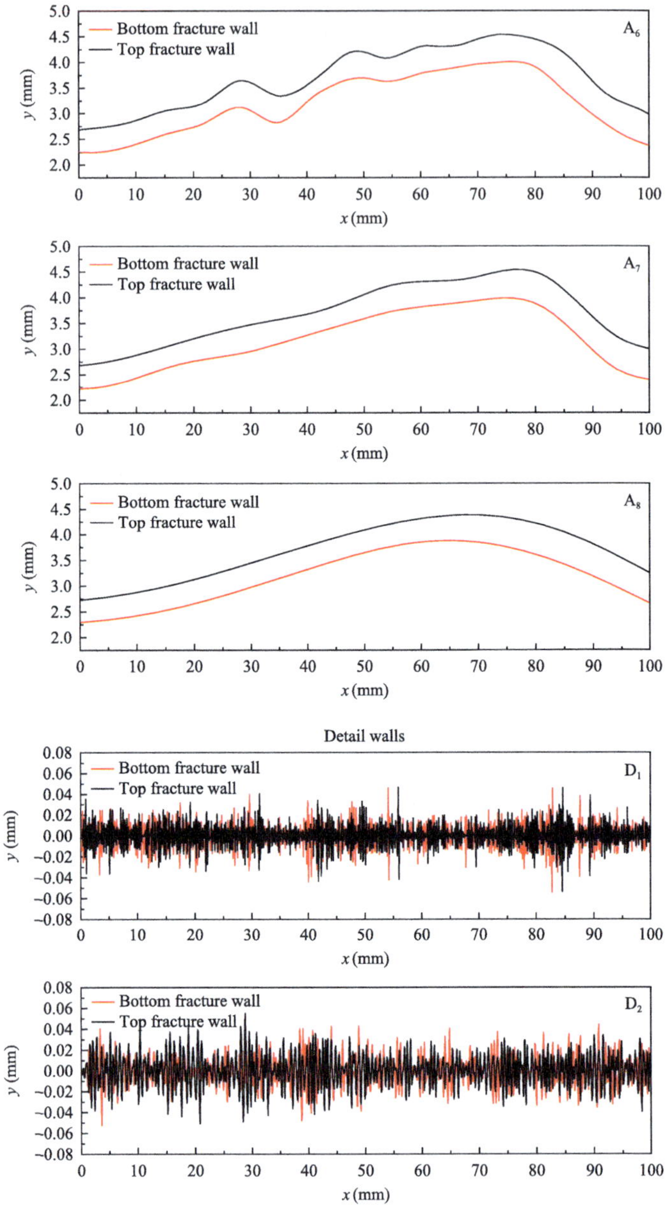

Fig. 9.3 (continued)

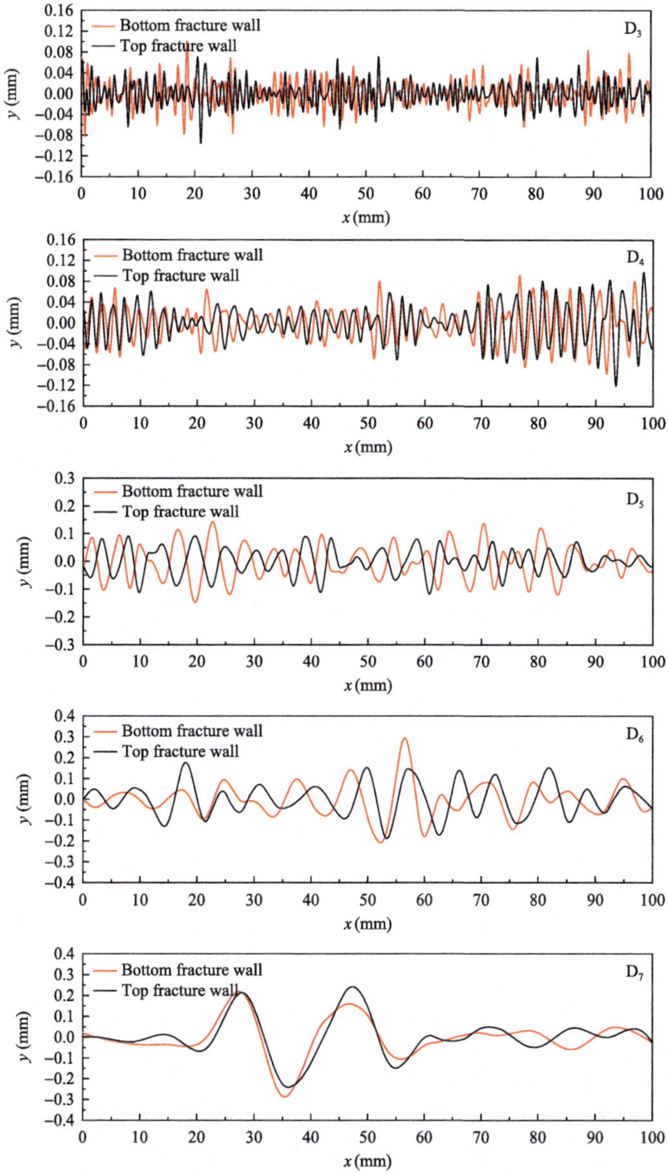

Fig. 9.3 (continued)

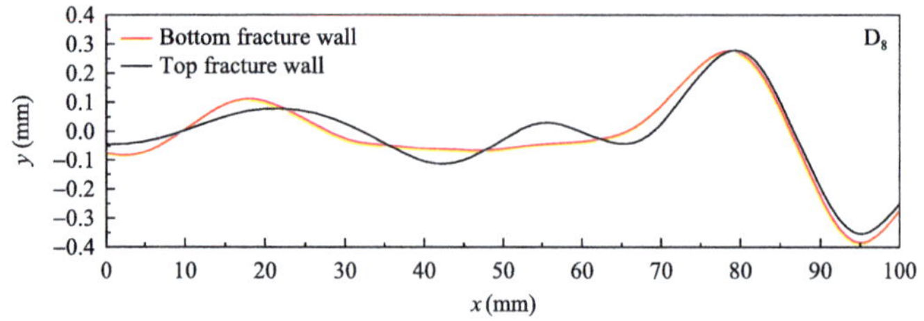

Fig. 9.3 (continued)

9.3 Flow Field Characteristics in Fractures

The flow field is highly dependent on the solute transport behavior. A nonlinear flow resulting in the different relationship between the D_L and *Pe* has been reported by Yeo and Ge (2001). Qian et al. (2011) conducted a well-controlled experiment to study the solute transport in a filled single fracture under non-Darcian (non-linear) flow. They found that the non-Darcian flow leads to a non-Fickian transport behavior and the dispersion under non-Darcian flow is much stronger than that under Darcian (linear) flow. Hence, it is necessary to state that the flow regime is controlled under the Darcian flow. For the flow field in fracture, the Reynolds number (*Re*) is defined as a the ratio of inertial forces to viscous forces,

$$Re = \frac{\rho Q}{\mu W} \tag{9.3.1}$$

where Q is the flow rate and W is the width of fracture in the out of plane direction (equal to 1 m in the 2D problem). Forchheimer's Law (Jacob, 1972) was given by

$$-\Delta p = aQ + bQ^2 \tag{9.3.2}$$

where $a = 12\mu/(We_h^3)$ and $b = \beta\mu/(W^2 e_h^2)$ are the linear and nonlinear coefficients, respectively, and β and e_h are the Forchheimer coefficient and hydraulic aperture, respectively.

When the inertial effect of fluid flow becomes negligible at low velocity, the nonlinear term [the second term on the right side of Eq. (9.3.2)] tends to be zero and the Forchheimer's Law reduces to the linear Darcy's Law.

The critical Reynolds number, Re_c, is proposed for characterizing the onset of flow transition to nonlinearity and given by

$$Re_c = \frac{a\rho\alpha}{b\mu W(1-\alpha)} \tag{9.3.3}$$

where α is the nonlinear effect factor defined as the ratio of the nonlinear pressure drop to the overall pressure drop

Table 9.1 Best-fitted results from Forchheimer equation and critical Re_c

Fracture	a [(kg/(s·m^5)]	b (kg/m^8)	Re_c
Original $H = 0.6$	4.54×10^7	1.77×10^{11}	13.8
Original $H = 0.7$	4.21×10^7	1.26×10^{11}	17.5
Original $H = 0.8$	4.14×10^7	7.81×10^{10}	27.8
Primary $H = 0.6$	3.66×10^7	4.77×10^{10}	40.2
Primary $H = 0.7$	3.44×10^7	4.07×10^{10}	44.2
Primary $H = 0.8$	2.92×10^7	3.27×10^{10}	46.8

$$\alpha = \frac{bQ^2}{aQ + bQ^2} \tag{9.3.4}$$

Table 9.1 shows that both the parameters a and b decrease as the Hurst exponent increases. For the same Hurst exponent, both of them decrease much more sharply in the original fracture than in the primary fracture where the secondary roughness is removed. As a result, the Re_c increases with the Hurst exponent increases for both original and primary fractures.

When the secondary roughness is removed, the corresponding Re_c increases significantly. This indicates that the secondary roughness has a considerable influence on nonlinear flow behaviors. This result shows a good agreement with the study of Wang et al. (2016). Here, we only considered the solute transport under Darican flow, which means that the Re was smaller than Re_c.

9.4 Relationship Between Tracer Longitudinal Dispersion and Peclet Number

1. Power-Law Relationship: $D_L/D_m \propto Pe^2$

The seminal Taylor's work (Aris, 1956)demonstrates that for solute transport through parallel plates where the fluid follows the stratified flow, the effective solute transport eventually becomes one dimensional and the shear-flow-induced dispersion can be measured by an effective longitudinal dispersion (i.e. Taylor dispersion) and mean flow velocity. The Taylor dispersion coefficient for an ideal 2D paralleled fracture is given as,

$$D_{\text{Taylor}} = \frac{\overline{u}^2\overline{b}^2}{210D_m} = \frac{Q^2}{210D_m} \tag{9.4.1}$$

Non-dimensionalizing the D_{Taylor} in terms of Pe reveals the power-law relationship between the D_{Taylor} and Pe,

$$\frac{D_{\text{Taylor}}}{D_{\text{m}}} = \frac{Pe^2}{210} \tag{9.4.2}$$

Depending on Eq. (9.4.1), Taylor dispersion coefficient can be directly estimated by the Q and D_{m} without conducting solute transport simulations.

Although previous studies have shown that the Taylor dispersion is expected to be valid for estimating D_{L} when *Pe* is relatively high so that the macro-dispersion and molecular diffusion can be negligible (Roux et al. 1998; Detwiler et al. 2000). It should be mentioned that even in an ideal 2D paralleled fracture, this is only valid at the asymptotic time scale and the length scale (Dentz and Carrera 2007; Bolster et al. 2009; Wang et al. 2012).

Based on the potential asymptotic time ($t = \overline{b}^2/4D_{\text{m}}$) in the 2D paralleled fracture, the critical *Pe* is calculated for each variable-aperture fracture. Below the critical *Pe*, the transport time in the given length scale is longer than the potential asymptotic time. This enables us to analyze the independent influence of multi-scale roughness on the relationship between the D_{L} and *Pe*, rather than the combined influence of multi-scale roughness and length scale. The same range of *Pe* is used to analyze the power-law, second-order, and linear relationship between the D_{L} and *Pe*.

In order to investigate the accuracy of Taylor dispersion in estimating the D_{L}, we proposed a dimensionless term $(D_{\text{Taylor}} + D_{\text{m}} - D_{\text{L}})/D_{\text{L}}$ as a metric. It can be seen in Fig. 9.4 that for all variable-aperture fractures, D_{Taylor} underestimates D_{L} at relatively low *Pe* while it overestimates D_{L} at relatively high *Pe*, which is in a good agreement with the previous result (Wang and Cardenas 2014). This indicates that the inaccurate D_{L} estimate from Taylor dispersion results from the roughness of the fracture, since the solute transport in the variable-aperture fractures is asymptotic. More precisely, our result shows that removing the secondary roughness of the fracture cannot remarkably improve accuracy of Taylor dispersion theory in estimating D_{L}.

The same tendency of $(D_{\text{Taylor}} + D_{\text{m}} - D_{\text{L}})/D_{\text{L}}$ against *Pe* is observed in both original and primary fractures, as is shown in Fig. 9.4. Furthermore, the value of

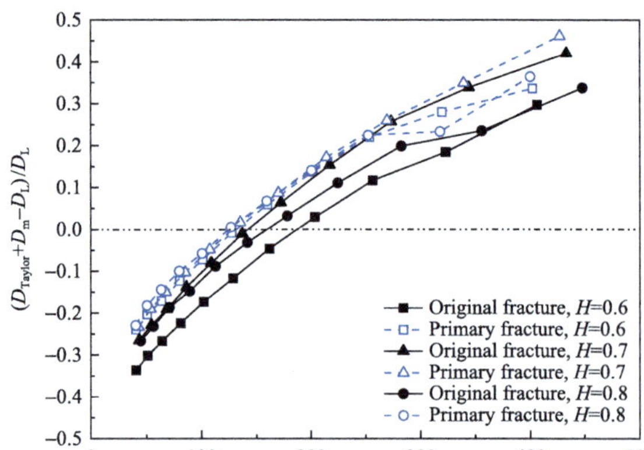

Fig. 9.4 Comparison of $(D_{\text{Taylor}} + D_{\text{m}} - D_{\text{L}})/D_{\text{L}}$ versus *Pe* in variable-aperture fractures

$(D_{Taylor} + D_m - D_L)/D_L$ varies slightly with the Hurst exponent in the primary fractures where the secondary roughness is not considered, but varies remarkably with the Hurst exponent in the original fractures. This implies that the inaccuracy of the D_L estimate from Taylor dispersion is not only due to the intensity of the heterogeneity of the roughness(i.e. Hurst exponent and COVs), but also due to the scale of the heterogeneity (i.e. primary and secondary roughness).

For an ideal 2D paralleled fracture, Taylor dispersion has the power-law relationship with *Pe* with an exponent equal to 2. Therefore, the exponent of the power-law relationship is interpreted as indicative of the influence of roughness on the D_L for the asymptotic solute transport.

As a result, although the relationship between the D_L/D_m and *Pe* exhibits a power law scaling, as shown in Fig. 9.5, the exponent of this power-law scaling is less than 2 for all the variable-aperture fractures. This also can be considered as a result from the influence of the macro-dispersion which is proportionally related to *Pe*.

In addition, it is found that removing the secondary roughness cannot recover the exponent to 2 but rather leads to the slight increase of the exponent for all the original fractures with the Hurst exponent of 0.6–0.8. This is due to the fact that the large-scale (primary) roughness still contributes to the macro-dispersion after the decomposition of the original fracture but the small-scale (secondary) roughness has a significant influence on the macro-dispersion.

2. Second-Order Polynomial Relationship: $D_L/D_m \propto \tau + Pe + Pe^2$

The complete dispersion mechanism in the variable-aperture fracture is defined as the sum of three mixing and spreading mechanisms, i.e., molecular diffusion, marco-dispersion (Detwiler et al. 2000; Plouraboué et al. 1998; Roux et al. 1998), and Taylor dispersion, which is given as,

$$D_L = \tau D_m + D_{macro} + D_{Taylor} \tag{9.4.3}$$

where τ is the tortuosity for diffusion in fractures and D_{macro} is the marco-dispersion. In this case, the longitudinal dispersion can be considered as the second-order polynomial relationship to *Pe*,

$$D_L/D_m = \tau + \alpha_{macro} Pe + \alpha_{Taylor} Pe^2 \tag{9.4.4}$$

where the dimensionless coefficients, α_{macro} and α_{Taylor}, represent the contributions of Taylor dispersion and macro-dispersion mechanisms, respectively. Thus, Eq. (9.4.4) provides insight into the *Pe* dependence of the mixing and spreading mechanism.

To investigate the influence of the multi-scale roughness on the second-order polynomial relationship between the D_L/D_m and *Pe*, we estimate the α_{macro} and α_{Taylor} by fitting the numerically generated relationship between the D_L/D_m and *Pe* in both primary and original fractures.

Table 9.2 shows the best-fitting results from the second-order polynomial relationship in Eq. (9.4.4). It can be seen that for all of the variable-aperture fractures,

Fig. 9.5 The power law relationship between the *Pe* and the ratio of D_L to the molecular diffusion D_m for the original and primary fractures with $H = 0.6$ (**a**), 0.7 (**b**), and 0.8 (**c**)

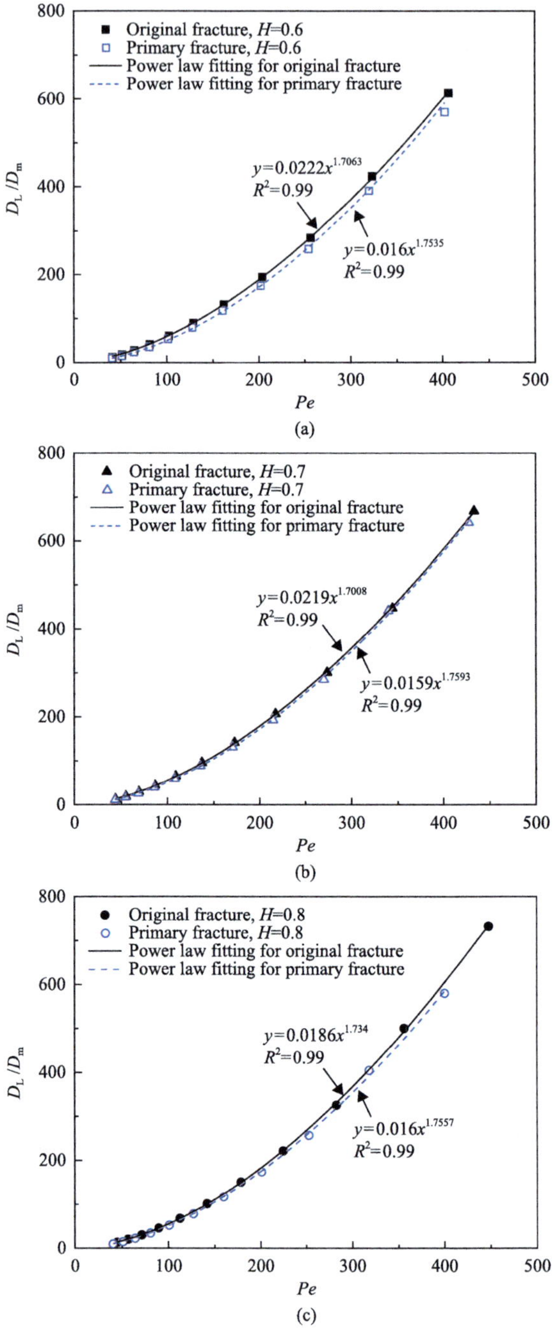

the α_{Taylor} is around 0.003 which is below the theoretical value of 0.00476(1/210)for an ideal 2D paralleled fracture. This result suggests that ① both the Hurst exponent and secondary roughness have a negligible influence on α_{Taylor} and ② other mixing and spreading mechanisms (i.e., diffusion and macro-dispersion) contribute the D_{L}. Moreover, removing the secondary roughness cannot result in a considerable change in α_{Taylor}. This reflects that the primary roughness, which is the large-scale roughness, is a dominant factor for the Taylor dispersion mechanism and leads to the value of α_{Taylor} less than the theoretical value. Since the COVs and $\overline{b}$ between the original and primary fractures with $H = 0.6$, 0.7, and 0.8 are almost constant, we postulate that the α_{Taylor} could be mainly depended on COVs and $\overline{b}$ rather than the secondary roughness.

As for the macro-dispersion mechanism, it is found in Table 9.2 that both the Hurst exponent and the secondary roughness have considerable influence on the α_{macro}. The α_{macro} decreases with the increase of the Hurst exponent in both original and primary fractures. This is due to the fact that the lower Hurst exponent, associating with the lower spatial correlation and the rougher wall, enhances the velocity variation. Detwiler et al. (2000) stated that macro-dispersion could be expected to dominate solute transport regime when $1/\alpha_{\text{macro}} < Pe < \alpha_{\text{macro}}/\alpha_{\text{Taylor}}$. Due to the almost constant value of α_{Taylor} among all the primary and original fractures, the *Pe* range where macro- dispersion dominates is only dependent on the α_{macro}. The *Pe* range decreases as the Hurst exponent increases no matter whether the secondary roughness is considered or not.

Since the macro-dispersion results from the velocity variation (Roux et al. 1998; Detwiler et al. 2000), one can expect that such enhanced velocity variation by decreasing Hurst exponent gives a rise to the increase of the *Pe* range where the macro-dispersion dominates. More importantly, α_{macro} decreases significantly as the secondary roughness is removed for the fractures with the same Hurst exponent. It can be found that removing the secondary roughness results in a larger decrease than increasing the Hurst exponent does. This implies that the secondary roughness has a more significant influence on the α_{macro}. In contrast to the primary (large-scale) roughness, the secondary (small-scale) roughness is a dominant factor for the macro-dispersion mechanism.

Table 9.2 Best-fitting results from the second-order polynomial relationship between the D_{L} and *Pe* in Eq. (9.4.4)

Fracture	α_{Taylor}	α_{macro}	*Pe* range where the macro-dispersion dominates
Original $H = 0.6$	0.0029	0.3332	$3.0 < Pe < 114.9$
Original $H = 0.7$	0.0029	0.3073	$3.3 < Pe < 106.0$
Original $H = 0.8$	0.0030	0.2893	$3.5 < Pe < 96.4$
Primary $H = 0.6$	0.0031	0.2724	$3.7 < Pe < 87.9$
Primary $H = 0.7$	0.0032	0.2659	$3.8 < Pe < 83.1$
Primary $H = 0.8$	0.0031	0.2471	$4.0 < Pe < 79.7$

3. Linear Relationship: $D_L/D_m \propto Pe$

It has been widely-recognized that the solute transport model can be conducted by a ID ADE with an effective longitudinal dispersion due to the absence of detailed information on the fracture aperture distribution. In this case, the D_L was assumed as the linear relationship to the mean velocity,

$$D_L^* = \alpha_L \bar{u} + D_m \tag{9.4.5}$$

In Eq. (9.4.5), α_L refers to the longitudinal dispersivity. Since the relationship between the D_L/D_m and *Pe* is assumed to be linear, α_L is expected to be linearly dependent on *Pe*. Notice that the α_L should not be confused with the dimensionless coefficient α_{macro} in Eq. (9.4.4). α_L can be considered as a result of simply lumping the contributions of the Taylor dispersion and macro-dispersion mechanisms, rather than the single contribution of the macro-dispersion mechanism.

Figure 9.6 shows the linear relationship between α_L and *Pe* for all the variable-aperture fractures. In general, α_L increases with *Pe* and is linearly related to *Pe*. However, it can be found that for *Pe* > 450 where the Taylor dispersion mechanism is dominant depending on Eq. (9.4.4), this linear relationship becomes weak. This indicates that the linear relationship between the D_L/D_m and *Pe* is inappropriate over a wide range of *Pe* where the different dispersion mechanism can be dominant.

Although the influence of the Hurst exponent and secondary roughness is confirmed in the power-law and second-order polynomial relationships, no distinct influence on this linear relationship can be observed. This reflects that the linear relationship in Eq. (9.4.5) is insensitive to the influence of multi-scale roughness in the variable-aperture fracture as the Taylor-dispersion mechanism dominates.

According to the Eq. (9.4.4), the dominance of each dispersion mechanism is dependent on the range of *Pe* and exhibits the different relationships to *Pe*. Since both α_L and α_{macro} are linearly related to *Pe*, one can expect that the linear relationship

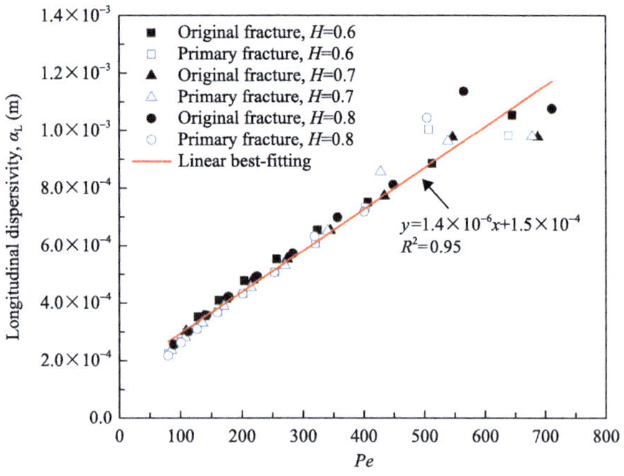

Fig. 9.6 The relationship between α_L and *Pe* in the variable-aperture fractures

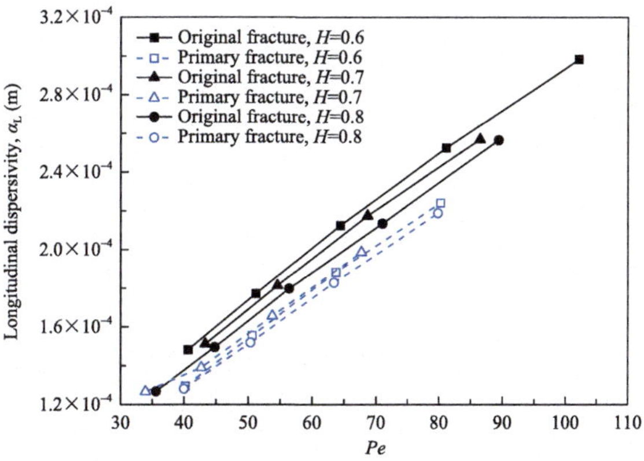

Fig. 9.7 The relationship between α_L and *Pe* in variable-aperture fractures over a range of *Pe* associated with dominance of the marco-dispersion

in Eq. (9.4.5) shows the better informative performance in the macro-dispersion-dominant range of *Pe* than in Taylor-dispersion-dominant range of *Pe*.

Figure 9.7 shows the influences of the Hurst exponent and secondary roughness on α_L in the range of *Pe* where the macro-dispersion is dominant. It can be seen that α_L is larger in the original fracture than in the primary fracture where the secondary roughness is removed. This indicates that the secondary roughness plays a significant role in enhancing α_L when the macro-dispersion mechanism is dominant. In addition, in the range of *Pe* where the macro-dispersion is dominant, the Hurst exponent is also found to enhance the α_L with and without consideration of the secondary roughness, associating that the larger Hurst exponent the smaller α_L.

References

Aris R (1956) On the dispersion of a solute in a fluid flowing through a tube. Proceed Roy Soc London a: Mathematic Phys Eng Sci 235(1200):67–77

Bolster D, Dentz M, Le Borgne T (2009) Solute dispersion in channels with periodically varying apertures. Phys Fluids 21(5):056601

Dentz M, Carrera J (2007) Mixing and spreading in stratified flow. Phys Fluids 19(1):017107

Detwiler RL, Rajaram H, Glass RJ (2000) Solute transport in variable-aperture fractures: an investigation of the relative importance of Taylor dispersion and macro-dispersion. Water Resour Res 36(7):1611–1625

Develi K, Babadagli T (1998) Quantification of natural fracture surfaces using fractal geometry. Math Geol 30(8):971–998

Dou Z, Zhou Z, Sleep BE (2013) Influence of wettability on interfacial area during immiscible liquid invasion into a 3D self-affine rough fracture: Lattice Boltzmann simulations. Adv Water Resour 61:1–11

Jacob B (1972) Dynamics of fluids in porous media. Elsevier, Amsterdam

Mallat SG (1989) A theory for multiresolution signal decomposition: the wavelet representation. IEEE Trans Pattern Anal Mach Intell 11(7):674–693

Mandelbrot BB (1983) The fractal geometry of nature. Am J Phys 51(3):286

Odling N (1994) Natural fracture profiles, fractal dimension and joint roughness coefficients. Rock Mech Rock Eng 27(3):135–153

Plouraboué F, Hulin JP, Roux S et al (1998) Numerical study of geometrical dispersion in self-affine rough fractures. Phys Rev E 58(3):3334–3346

Qian J, Chen Z, Zhan HB et al (2011) Solute transport in a filled single fracture under non-Darcian flow. Int J Rock Mech Min Sci 48(1):132–140

Roux S, Plouraboué F, Hulin J-P (1998) Tracer dispersion in rough open cracks. Transp Porous Media 32(1):97–116

Wang L, Cardenas MB (2014) Non-Fickian transport through two-dimensional rough fractures: assessment and prediction. Water Resour Res 50(2):871–884

Wang JSY, Narasimhan TN, Scholz CH (1988) Aperture correlation of a fractal fracture. J Geophys Res Solid Earth 93(B3):2216–2224

Wang L, Cardenas MB, Deng W et al (2012) Theory for dynamic longitudinal dispersion in fractures and rivers with poiseuille flow. Geophys Res Lett 5:L05401

Wang L, Cardenas MB, Slottke DT et al (2015) Modification of the local cubic law of fracture flow for weak inertia, tortuosity, and roughness. Water Resour Res 51(4):2064–2080

Wang M, Chen YF, Ma GW et al (2016) Influence of surface roughness on nonlinear flow behaviors in 3D self-affine rough fractures: Lattice Boltzmann simulations. Adv Water Resour 96:373–388

Yeo IW, Ge S (2001) Solute dispersion in rock fractures by non-Darcian flow. Geophys Res Lett 28(20):3983–3986

Zimmerman R, Kumar S, Bodvarsson G (1991) Lubrication theory analysis of the permeability of rough-walled fractures. Int J Rock Mech Mining Sci Geomech Abstr 28(4):325–331

Zou L, Jing L, Cvetkovic V (2015) Roughness decomposition and nonlinear fluid flow in a single rock fracture. Int J Rock Mech Min Sci 75:102–118

Chapter 10
Influence of Eddies on Solute Transport Through a Fracture

The flow field's heterogeneity has a big impact on the solute transport regime. In this chapter, Navier–Stokes flow and solute transport simulations are used to examine the influence of eddies on conservative solute transport through a two-dimensional (2D) single self-affine fracture.

10.1 Flow Field and Eddies Formation

Fickian tendencies are typically considered to govern solute transport. The well-known advection–dispersion equation (ADE) (Bear 1972) is the governing equation for Fickian transport, and it is predicated on the assumption that the dispersion coefficient is constant in both space and time. Gaussian-shaped BTCs can be foreseen with high accuracy via the Fickian ADE. Nevertheless, in-depth field and experiment studies (Gouze et al. 2008; Zhang et al. 2007; Dou and Zhou 2015) have revealed that BTCs are not always Gaussian and that the Fickian transport model is insufficient to describe non-Gaussian-shaped BTCs.

Two common traits can be found in non-Gaussian-shaped BTCs for a pulse plume. Secondly, the concentration peak in BTCs arrived earlier than the Fickian transport model expected, indicating that the solute transport may have taken a preferential path (Kosakowski et al. 2001). The alleged "early arrival" phenomenon is this. Second, compared to what the Fickian transport model predicted, the tail that follows the concentration peak in BTCs lasts longer. This "heavy tail" may be explained by a multi-rate mass exchange mechanism between mobile and immobile domains (Haggerty and Gorelick 1995; Cvetkovic and Haggerty 2002), the absorption of the fracture wall (Zimmerman et al. 2002; Bodin et al. 2003), the permeable rock matrix (Zhou et al. 2006), the presence of eddies inside the fracture (Cardenas et al. 2007, 2009; Lee et al. 2015), and other factors.

Z. Dou et al., *Mass Transfer Dynamics of Contaminants in Fractured Media*,
https://doi.org/10.1007/978-981-99-9187-7_10

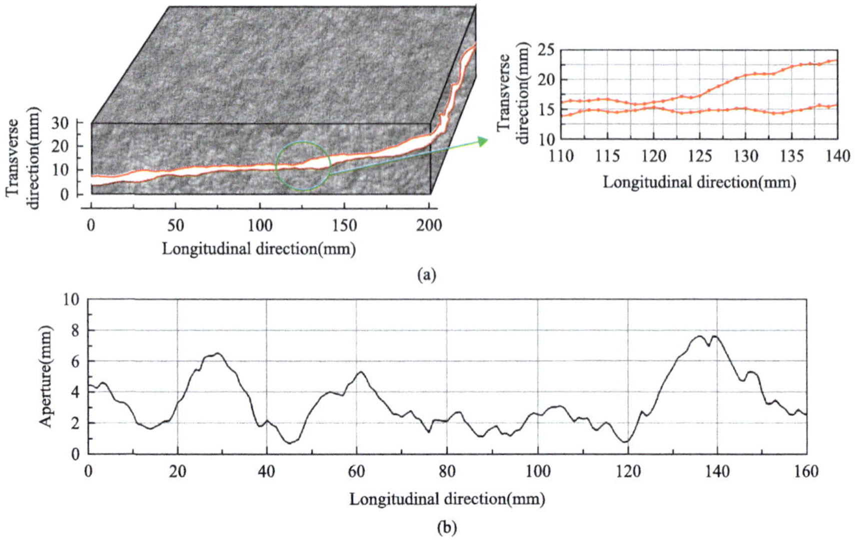

Fig. 10.1 Generated single self-affine fracture with $H = 0.8$ by using the successive random additions technique **a** Illustration of the generated self-affine fracture in rocks; **b** local aperture as a function of longitudinal distance along the fracture

Although the non-Fickian solute transport has been studied in different research areas such as heat conduction or mass transport (Gómez et al. 2007), fast crystallization process (Galenko et al. 2015), percolation through the porous media (Gouze et al. 2008), and diffusion in solid polymers (Kalospiros et al. 1991), we focus on the influence of eddies on solute transport through a single self-affine fracture.

Figure 10.1 illustrates the generation of a single self-affine facture with H = 0.8 using the successive random additions method.

Studying the conservative solute transport requires a fundamental understanding of the flow field in fractures. The employed flow fields are introduced and briefly described for completeness' sake. We employ four different pressure gradients to drive the flow field in a single self-affine fracture because the occurrence of eddies is very sensitive to the properties of the flow field. As a result, four distinct flow fields (denoted by FF1–FF4) are obtained. In order to quantify the features of the flow field, the Reynolds number (Re) is introduced. The definition of Re is the ratio of inertial to viscous forces,

$$Re = \frac{\rho Q}{\mu W} \tag{10.1}$$

where Q is the flow rate and W is the width of fracture in the out of plane direction (equal to 1 m in the 2D problem).

A segment of the 160 mm fracture with a length of 22 mm and a Hurst exponent of 0.8 is shown in Fig. 10.2. There is no eddy in the segment at low $Re (Re = 30)$. In actuality, the flow field demonstrates that at the full-length scale of the fracture, the effective advection zone completely fills all of the cross-sectional regions and there is no discernible eddy region. Eddies gradually occupy a large section of the fracture as *Re* rises, especially where the aperture expands suddenly. The growth of the eddy areas causes the effective advection region to become smaller. Similar findings for the expansion of eddies due to the increasing *Re* are shown in a number of earlier publications (Bouquain et al. 2012; Qian et al. 2012; Zou et al. 2015; Briggs et al. 2016; Wang et al. 2016). It is obvious that for high *Re* ($Re = 170$), eddies regions occupy a bigger fraction of the fracture area and the main flow channel becomes smoother, even if the effective advection region is smaller than other situations with lower *Re*.

The streamlines can separate from the bulk flow due to the existence of eddies and then re-engage at the downstream area (Armaly et al. 1983). As a result, bulk flow is not affected by the detached flow areas. Moreover, the streamlines meander and potentially change directions in the eddy region because to the extremely low velocity. Given that the local *Pe* is modest and molecular diffusion takes over as the

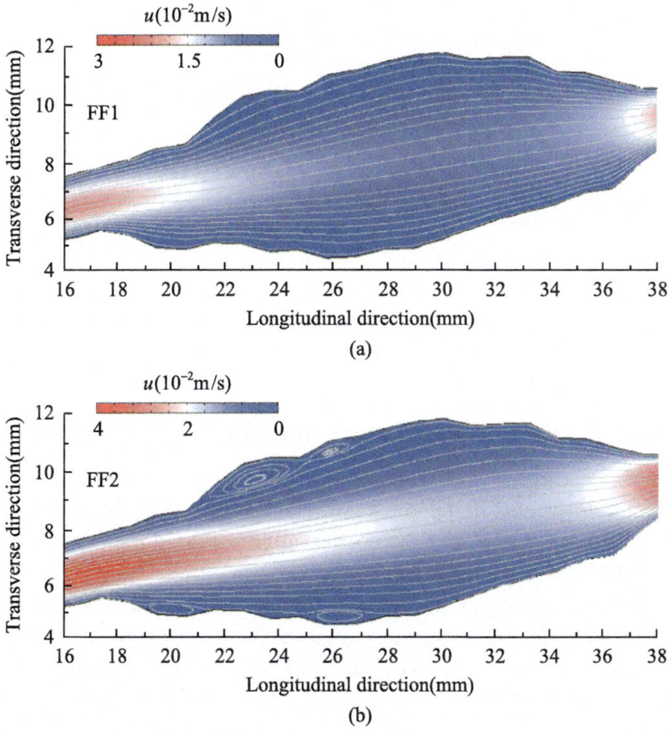

Fig. 10.2 The segments of the flow field (FF1–FF4) in the single self-affine fracture with $Re = 30$, 60, 100, and 170 **a** $Re = 30$; **b** $Re = 60$; **c** $Re = 100$; **d** $Re = 170$

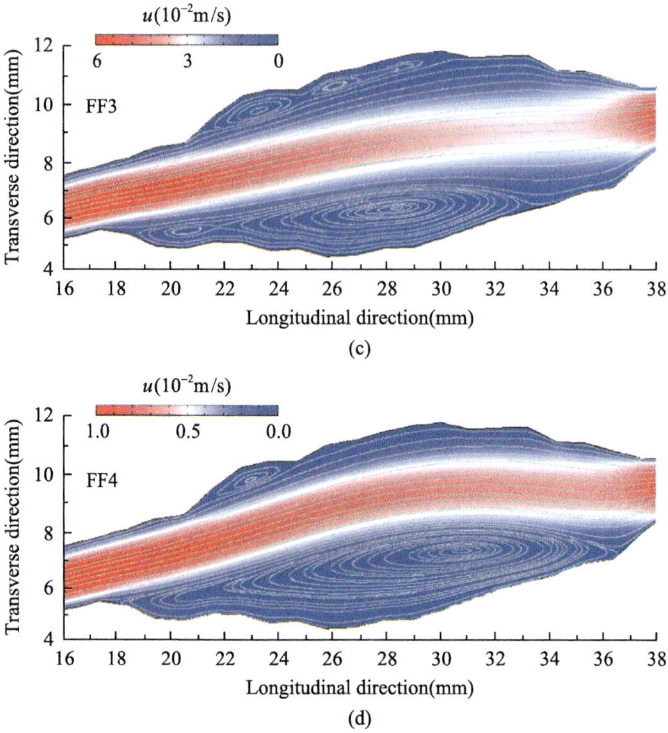

Fig. 10.2 (continued)

dominant mechanism (Cardenas et al. 2007), this could cause solute transport to be delayed.

10.2 Spatial Evolution of Solute and BTC Characteristics

For each of the flow fields in the crack, simulations of conservative solute transport are run. The spatial evolution of the conservative solute in the crack with the flow fields FF1 and FF4 is depicted in Fig. 10.2. The same pore volumes (i.e., PV = 0.2, 0.4, 0.8, and 1.2) in Fig. 10.3 are selected for FF1 and FF4 in order to examine the impact of aperture and eddies on the spatial evolution of the conservative solute. It should be noted that due to the continually injected solute, the conservative solute distribution should be uniform at the late time. Figure 10.3 illustrates how the spatial evolution of the solute is dependent on the distribution of apertures. In comparison to the region close to the fracture walls, the solute moves more quickly in the center of the fracture. It demonstrates unequivocally that the fracture has a preferred channel for the transport of solutes in the longitudinal direction. The spatial evolution of

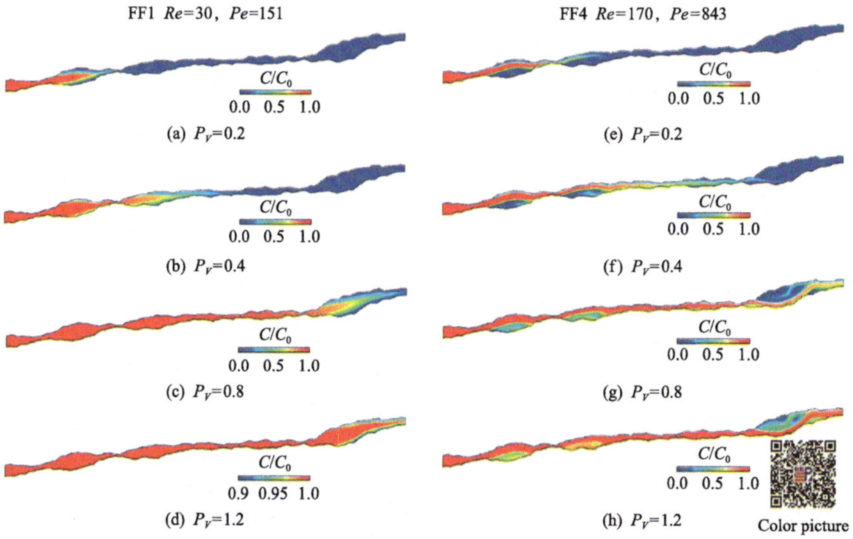

Fig. 10.3 The spatial evolution of the plume in the single self-affine fracture with the different flow fields (**a**–**d**) in the FF1 and (**e**–**h**) in the FF4

the solute is slightly delayed in large-aperture locations for the relatively small *Re*. Due to the non-slip boundary condition of fracture walls, the flow velocity at the fracture wall is significantly lower than that in the middle zone, which may help to explain this. Diffusion hence dominates transport near fracture walls, in contrast to the advection-dominated transport in the main flow channel.

Eddies grow in places with relatively big apertures as *Re* rises. We can draw the conclusion that the solute first passes past eddies without their involvement and then typically becomes caught in eddies downstream. As soon as eddies catch the solute from the bulk flow, it starts to spin and slowly move back upstream (see Fig. 10.3f, g). Due to the tiny velocity compared to the bulk velocity, the diffusion mechanism may still predominate the spreading and mixing processes in eddies, but these eddies create a momentary "immobile" zone that can be very resistant to spreading and mixing. Eddies will therefore cause a significant delay in the solute's delivery.

For FF1–FF4, the relevant flux-weighted BTCs at the fracture's exit are determined (Fig. 10.4). It is extremely difficult to characterize the dilution or mixing of the solute by integrated measurements of solute breakthrough because the spreading and mixing processes that regulate the spatial evolution of the solute are highly coupled with one another and dependent on the flow fields during the solute transport (Chiogna et al. 2011; de Dreuzy et al. 2012; Rolle and Kitanidis 2014). Here our focus is not to evaluate dilution processes by BTCs. As an alternative, we stress that the presence of eddies in the flow field has a substantial impact on solute transport and associated BTCs, which can be significant factors in the delay of solute transport and the increase of non-Fickian transport features. All the BTCs exhibit the typical non-Fickian traits, such as "early arrival" and "heavy tail". An ideal Fickian

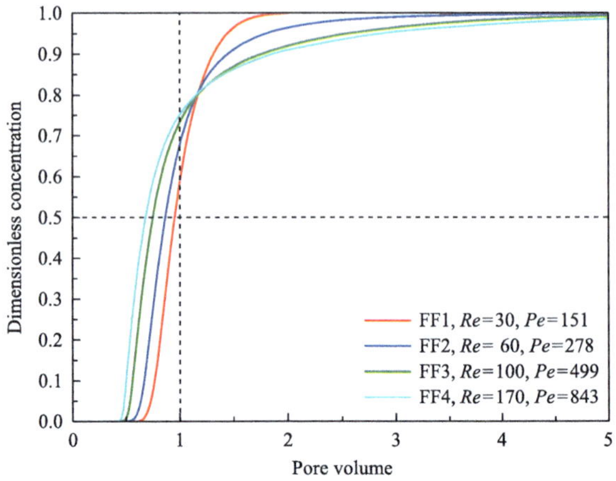

Fig. 10.4 Breakthrough curves at the outlet of the single self-affine fracture with different *Re* and *Pe*

transport resulting to a typical Gaussian-shaped plume under the step injection condition should demonstrate that the dimensionless concentration of BTCs is equal to 0.5 following a single P_V injection. The results demonstrate that the dimensionless concentrations at $P_V = 1$ for all BTCs are more than 0.5, indicating "early arrival" of the plume with or without eddies. In actuality, there is no eddy in the lone self-affine fracture for the small $Re(Re = 30)$. The fundamental causes of the "early arrival" in BTCs in this instance are the convergence and divergence of streamlines brought about by the abrupt enlargement of the aperture. A direct consequence of the improved spreading (dispersion) process in the primary flow channel is that as Re rises, the "early arrival" tends to be enhanced. The diffusion contributes very little to the "early arrival" compared to the advection-dominated transport in the main flow channel, regardless of whether eddies appear in the flow field or not.

The "heavy tail" of BTCs can be created by eddies that form in the rough cracks (Cardenas et al. 2007; Lee et al. 2015). The residence time distributions (RTDs) are computed for the solute transport under various four flow fields in order to describe the characteristics of the "heavy tail" of BTCs. RTDs at the outlet can readily be derived from the time derivative of BTCs because solving the Eulerian advection–diffusion-equation with step injection results in the resident concentration time series,

$$R(t) = \frac{\mathrm{d}c(t)}{\mathrm{d}t} \tag{10.2}$$

The RTDs for the solute transport through the solitary self-affine fracture under FF1–FF4 are shown in Fig. 10.5. Instead of the power-law tail, the RTDs should display a dramatic Fickian drop-off after the apex if the solute transport complies with the Fickian law (Cardenas et al. 2007, 2009). Figure 10.5 makes it abundantly obvious that RTD tails in FF1–FF4 exist whether or not eddies are present. Since there is not an eddy that can be seen in FF1, the varying aperture causes the RTD tails. The heavier tails can be spotted in FF2–FF4 as eddies appear and grow with *Re*. In

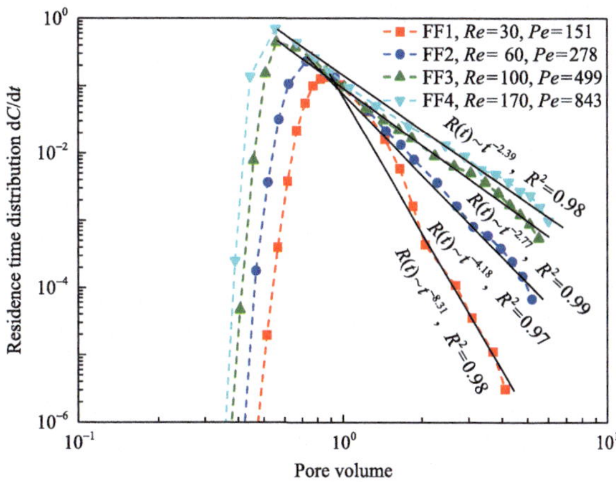

Fig. 10.5 Residence time distributions (RTDs) for the single self-affine fracture with different *Re* and *Pe*

the meantime, even after peaks, the RTDs diverge from the Fickian distribution. For all RTDs, we also fit the tails after the peaks using the power-law function of time. A power-law decline is discovered to be followed by the tails. According to the fitted exponent for a power-law function, the bigger eddy causes the heavier tail, as shown by the rise in fitted exponent with respect to *Re*. The backflow can actually trap the solute particles in eddies or recirculation areas of the flow field (see Fig. 10.3f, g). Eddies are thought of as stationary areas where the solute may engage in multiple rate exchange processes (Cvetkovic and Haggerty 2002). The size of the tails should be explained by the efficient rate of mass exchange. Even while we are unable to precisely measure the mass exchange processes between the main flow channel and eddies, our qualitative findings show that eddies are the origin of tails, and greater eddies are associated with heavier tails, which is consistent with the findings from (Boutt et al. 2006; Cardenas et al. 2007). The solute is initially captured in and swirls around in eddies as they form, Lee et al. (2015) describe, and then is thrown back into the main flow channel. The result of this is that the tail gets shorter. The discrepancy can be the result of the flow. The nonlinear flow behavior will be crucial in improving the mixing processes in eddies and the mass exchange rate between the main flow channel and eddies if *Re* is large enough. The phenomena of the shorter tails cannot be observed in our study because the nonlinear flow behavior is not clear enough.

10.3 Inverse Model for Non-Fickian BTCs

The solute transport is non-Fickian due to the presence of eddies in the crack, and the resulting BTCs display traits like "early arrival" and "heavy tail," which can be theoretically described by several transport models. Here, two alternative inverse models—the traditional 1D ADE and the 1D CTRW TPL models, respectively—fit the non-Fickian BTCs derived from the direct simulations. The convergence to a minimum error between the inverse model and simulated BTCs from the direct simulations will be aided by an accurate initial guess for fitting parameters (Wang and Cardenas 2014, 2015). We develop a global error metric, E_i, to measure the goodness-of-fit,

$$E_i = \sqrt{\sum_i^N (c_{\text{fit},i} - c_{\text{model}})^2/N} \tag{10.3}$$

where i is either the ADE or the CTRW TPL model, N is the number of the calculated data in the BTCs, $c_{\text{fit},i}$ constitutes the fitted BTCs using the ADE or CTRW TPL model, and c_{model} represents the flux-weighted, spatially integrated BTC from the direct simulations. Hereafter global errors from the ADE and CTRW TPL models are denoted as E_{ADE} and E_{CTRW}, respectively. For the inverse model ADE, the velocity and dispersion coefficient are considered as the fitted parameters. The fitting is implemented by the nonlinear fitting toolbox, 1stOpt (7D-Soft High Technology Inc., Beijing, China). As a result, the fitted dispersion D_{ADE} and the global error for the ADE inverse model E_{ADE} are obtained. For the inverse model with the CTRW TPL, there are five initial input parameters, t_1, t_2, u_0, β and D_0. The initial guess for velocity u_0 and dispersion D_0 is assumed to be equal to the values from the ADE inverse modeling. The CTRW Matlab Toolbox V3.1 (Cortis and Berkowitz 2005) is employed for the CTRW TPL inverse model.

The fitting outcomes for the ADE and CTRW TPL inverse models are shown in Table 10.1. It can be shown that the global error for the ADE and CTRW TPL inverse models rises with *Re*, indicating that the flow field with eddies affects both inverse models. Nonetheless, in all flow fields (FF1–FF4), the global error is bigger for the ADE inverse model than for the CTRW TPL model, showing that the CTRW TPL model is more capable of capturing the properties of the non-Fickian transport. For two inverse models, the fitted dispersion coefficients rise with *Re*. As the dispersion quantifies the rate of spreading of the solute plume, which does not always accurately reflect the genuine mixing process, this finding also demonstrates that when Re grows, the spreading process enhances the "early arrival" of the BTCs. Consequently, the fitted dispersion alone cannot fully capture the "heavy tail," a crucial aspect of non-Fickian transport that is extremely sensitive to the mixing process. This is also the reason why many studies emphasize the solute transport's mixing process and contend that the ADE is insufficient to adequately describe mixing or dilution. Also, every fitted parameter is less than 2 and decreases with Re, indicating that every solute transport is non-Fickian and that the non-Fickian traits are strengthened by the high Reynolds number. According to the fitting results, the parameter β can be used to detect the presence of eddy-induced non-Fickian transport.

We can anticipate that the eddy volume can be associated to the magnitude of the eddy-induced non-Fickian transport (i.e., the parameter in CTRW TPL model) since the growing Re causes the expansion of eddies (Fig. 10.3). Here, the eddy volume in the self-affine fracture is inferred using the idea of the effective hydraulic aperture (Zou et al. 2015; Briggs et al. 2016). Using the cubic law, which is given as, one can

Table 10.1 Fitted parameters and global errors for the ADE and CTRW TPL model

Flow field	1D ADE model		CTRW TPL model				
	D_{ADE} (m^2/s)	E_{ADE}	t_1 (s)	t_2 (s)	β	D_{CTRW} (m^2/s)	E_{CTRW}
FF1	3.67×10^{-5}	0.0022	$10^{-0.91}$	$10^{0.99}$	1.81	3.66×10^{-5}	0.0019
FF2	1.44×10^{-4}	0.0052	$10^{-2.05}$	$10^{2.49}$	1.64	8.66×10^{-5}	0.0027
FF3	6.16×10^{-4}	0.0092	$10^{-6.91}$	$10^{1.97}$	1.02	1.14×10^{-4}	0.0026
FF4	1.54×10^{-3}	0.0099	$10^{-6.68}$	$10^{1.15}$	0.91	3.37×10^{-4}	0.0032

determine the effective hydraulic aperture, $\overline{b}_e$,

$$\overline{b}_e = \left(\frac{12\mu QL}{-W\Delta p}\right)^{1/3} \tag{10.4}$$

For FF1, FF2, FF3, and FF4, $\overline{b}_e$ values are 3.11, 2.84, 2.13, and 1.99 mm, respectively, which are below the mean physical aperture ($\overline{b} = 3.35$ mm). The area of the crack that contributes to the effective advection flow is proportional to the effective hydraulic aperture as a percentage of the mean physical aperture. The eddy volume, which is thought of as a different portion of the mean physical aperture, contains complex flows and sinuous streamlines that cause the solute transport to be delayed. The ratio of the effective hydraulic aperture to the mean aperture can be used to correlate the ratio of the bulk flow volume to the total fracture volume, whereas the ratio of the eddy volume to the total fracture volume can be calculated by subtracting 1 from the ratio of the effective hydraulic aperture to the mean aperture. Thus, for FF1, FF2, FF3, and FF4, the eddy volume ratios are 7, 15, 36, and 41%, respectively, indicating that as *Re* increases the eddy volume ratio increases. This is in line with the qualitative findings in Fig. 10.2 quite well. This shows that the parameter β for the CTRW TPL model decreases as eddy volume increases.

10.4 Uniformity of Concentration Distribution

The existence of the tails with step injection suggests the presence of a fracture and a non-uniform concentration distribution simultaneously. The concentration is higher in the main flow channel than it is in the eddy area, as shown in Fig. 10.3. The main cause of the uneven concentration distribution is eddies. Due to concentration gradients from the continually injected solute, this non-uniform concentration distribution tends to be uniform. The effluent solute concentration is equal to the beginning concentration once the mixing has spread out the concentration gradient. The size of the tails of BTCs is determined by the evolution process from a non-uniform to a uniform concentration distribution; the longer the evolution phase, the heavier the tail. We employ Kitanidis (1994) original notion of the dilution index to quantify this evolution process. The exponential Shannon entropy of a concentration probability distribution is represented by the dilution index, which is defined as,

$$E(t) = \exp\left\{-\int_{-\infty}^{+\infty}\int_{-\infty}^{+\infty} p_g(x, y, t)\ln[p_g(x, y, t)]\mathrm{d}x\mathrm{d}y\right\} \tag{10.5}$$

where $p_g(x, y, t)$ is the concentration probability function of the solute and defined by

$$p_g(x, y, t) = \frac{c(x, y, t)}{\int_{-\infty}^{+\infty}\int_{-\infty}^{+\infty} c(x, y, t)\mathrm{d}x\mathrm{d}y} \tag{10.6}$$

The greatest value of the dilution index, $E(t)_{\max}$, for a 2D fracture, which symbolizes a uniform distribution of concentrations within the fracture, is theoretically equal to the area of the entire fracture. Dilution is defined as "the act of dispersing solute mass over a greater volume" in Eq. (10.5). It should be noted that the effective volume occupied by a solute plume serves as the illustrative explanation of the dilution index from Eq. (10.5). Undeniably, the Shannon entropy in Eq. (10.5) has the property that the total solute mass within the fracture is not conservative because of the effluent solute at the outlet boundary. This is a useful property for not only analyzing the solute dilution process when the solute is injected as a single slug, but also quantifying the uniformity of concentration distribution within the fracture. As a result, there is no indication of the actual mixing process in the temporal evolution of the dilution index from Eq. (10.5). As our focus is in examining the uniformity of concentration distribution within the fracture, the dilution index that is being studied here is still relevant. The bigger $E(t)$, the better the uniformity of concentration distribution within the fracture.

The temporal evolution of the dilution index in a single self-affine fracture with various *Re* and *Pe* is shown in Fig. 10.6. The area of the complete fracture is used to normalize the dilution index. The dimensionless dilution index for the tiny *Re*($Re = 30$) is equivalent to one P_V after another. However, the dimensionless dilution index for the big *Re*($Re = 170$) is less than one after one P_V, showing the presence of the fracture's non-uniform concentration distribution. The eddy is mostly to blame for the non-uniformity of the concentration distribution, as illustrated in Fig. 10.3. Early on ($P_V < 0.2$), the dilution index's temporal evolution is independent of the *Re* and *Pe*, demonstrating that the heterogeneous flow field has no impact on the uniformity of the concentration distribution, eddies present or not. The dilution index between $1.0 > P_V > 0.2$ is improved by the big *Re*. This is due to the fact that as the solute is repeatedly injected, eddies catch an increasing amount of it. Another factor is that the quick advection that drives the plume tends to increase concentration differences, which also tends to increase solute dilution. After $P_V > 1.0$, it becomes apparent that the dilution index for the big *Re* is less than one, indicating that the eddies are delaying the uniformity of concentration distribution within the crack. The presence of concentration gradients between the eddy region and the main flow channel suggests that eddies operate as a barrier to the movement of solute through the crack. Since the mass exchange between the eddy area and the main flow channel

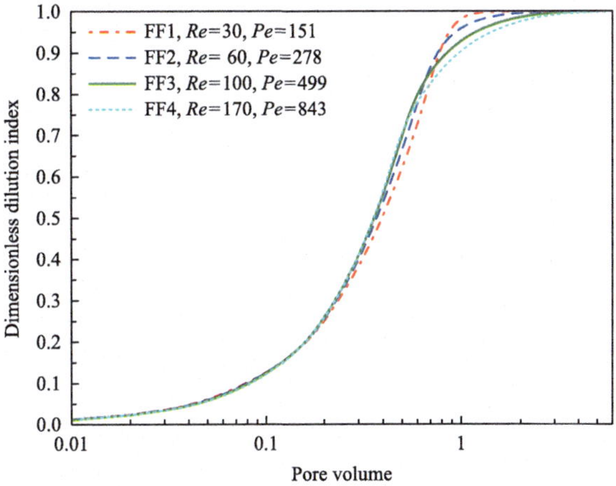

Fig. 10.6 Evolution of dilution index in a single self-affine fracture with the different *Re* and *Pe*

is what causes the tails of BTCs, the eddy is crucial in minimizing or delaying it. The delayed bulk exchange process determines the size of the BTC tails (Fig. 10.4).

References

Armaly BF, Durst F, Pereira J et al (1983) Experimental and theoretical investigation of backward-facing step flow. J Fluid Mech 127:473–496

Bear J (1972) Dynamics of fluids in porous materials. Society of Petroleum Engineers, Richardson

Bodin J, Delay F, de Marsily G (2003) Solute transport in a single fracture with negligible matrix permeability: 1. fundamental mechanisms. Hydrogeol J 11(4):418–433

Bouquain J, Méheust Y, Bolster D et al (2012) The impact of inertial effects on solute dispersion in a channel with periodically varying aperture. Phys Fluids 24(8):083602

Boutt DF, Grasselli G, Fredrich JT et al (2006) Trapping zones: The effect of fracture roughness on the directional anisotropy of fluid flow and colloid transport in a single fracture. Geophys Res Lett 33(21):L21402

Briggs S, Karney BW, Sleep BE (2016) Numerical modeling of the effects of roughness on flow and eddy formation in fractures. J Rock Mech Geotechn Eng 9(2017):105–115

Cardenas MB, Slottke DT, Ketcham RA et al (2007) Navier-Stokes flow and transport simulations using real fractures shows heavy tailing due to eddies. Geophys Res Lett 34(14):L14404

Cardenas MB, Slottke DT, Ketcham RA et al (2009) Effects of inertia and directionality on flow and transport in a rough asymmetric fracture. J Geophys Res 114(B6):B6204

Chiogna G, Cirpka OA, Grathwohl P et al (2011) Relevance of local compound-specific transverse dispersion for conservative and reactive mixing in heterogeneous porous media. Water Resour Res 47(7):W07540

Cortis A, Berkowitz B (2005) Computing "anomalous" contaminant transport in porous media: the CTRW MATLAB toolbox. Ground Water 43(6):947–950

Cvetkovic V, Haggerty R (2002) Transport with multiple-rate exchange in disordered media. Phys Rev E 65(5):051308

de Dreuzy JR, Carrera J, Dentz M et al (2012) Time evolution of mixing in heterogeneous porous media. Water Resour Res 48(6):W06511

Dou Z, Zhou ZF (2015) Modelling of solute transport in a filled fracture: effects of heterogeneity of filled medium. J Hydrodynam Ser B 27(1):85–92
Galenko PK, Danilov DA, Alexandrov DV (2015) Solute redistribution around crystal shapes growing under hyperbolic mass transport. Int J Heat Mass Transf 89(Supplement C:1054–1060.
Gómez H, Colominas I, Navarrina F et al (2007) A finite element formulation for a convection–diffusion equation based on Cattaneo's law. Comput Methods Appl Mech Eng 196(9):1757–1766
Gouze P, Le Borgne T, Leprovost R et al (2008) Non-Fickian dispersion in porous media: 1. Multiscale measurements using single-well injection withdrawal tracer tests. Water Resour Res 44(6):W06426
Haggerty R, Gorelick SM (1995) Multiple-rate mass transfer for modeling diffusion and surface reactions in media with pore-scale heterogeneity. Water Resour Res 10:2383–2400
Kalospiros NS, Ocone R, Astarita G et al (1991) Analysis of anomalous diffusion and relaxation in solid polymers. Ind Eng Chem Res 30(5):851–864
Kitanidis PK (1994) The concept of the dilution index. Water Resour Res 30(7):2011–2026
Kosakowski G, Berkowitz B, Scher H (2001) Analysis of field observations of tracer transport in a fractured till. J Contam Hydrol 47(1):29–51
Lee SH, Yeo IW, Lee KK et al (2015) Tail shortening with developing eddies in a rough-walled rock fracture. Geophys Res Lett 42(15):6340–6347
Qian J, Liang M, Chen Z et al (2012) Eddy correlations for water flow in a single fracture with abruptly changing aperture. Hydrol Process 26(22):3369–3377
Rolle M, Kitanidis PK (2014) Effects of compound-specific dilution on transient transport and solute breakthrough: a pore-scale analysis. Adv Water Resour 71:186–199
Wang L, Cardenas MB (2014) Non-Fickian transport through two-dimensional rough fractures: assessment and prediction. Water Resour Res 50(2):871–884
Wang L, Cardenas MB (2015) An efficient quasi-3D particle tracking-based approach for transport through fractures with application to dynamic dispersion calculation. J Contam Hydrol 179:47–54
Wang M, Chen YF, Ma GW et al (2016) Influence of surface roughness on nonlinear flow behaviors in 3D self-affine rough fractures: Lattice Boltzmann simulations. Adv Water Resour 96:373–388
Zhang Y, Benson DA, Baeumer B (2007) Predicting the tails of breakthrough curves in regional-scale alluvial systems. Groundwater 45(4):473–484
Zhou Q, Liu HH, Bodvarsson GS et al (2006) Evidence of multi-process matrix diffusion in a single fracture from a field tracer test. Transp Porous Media 63(3):473–487
Zimmerman MD, Bennett PC, Sharp JM et al (2002) Experimental determination of sorption in fractured flow systems. J Contamin Hydrol 58(1–2):51–77
Zou L, Jing L, Cvetkovic V (2015) Roughness decomposition and nonlinear fluid flow in a single rock fracture. Int J Rock Mech Min Sci 75:102–118

Chapter 11
Lattice Boltzmann Simulation of Solute Transport in Fractures

The lattice Boltzmann method (LBM) has gained considerable popularity in recent years. A number of complex fluid flow problems, including single- and multi-phase flow in complex geometries, can be computationally modeled using the LBM, which is a potent technique. Based on LBM, the purpose of this chapter is to examine how a rough fracture wall affects the solute transport in a single fracture.

11.1 Coupling Flow and Concentration Fields Based on LBM

The flow and concentration fields are represented, respectively, by two distribution functions. Each distribution function's evolution equation satisfies the lattice Boltzmann equation below with a single relaxation time:

$$f_i(X + e_i \Delta t, t + \Delta t) = f_i(X, t) - \frac{\Delta t}{\tau}\left[f_i(X, t) - f_{\mathrm{eq}i}(X, t)\right] \tag{11.1.1}$$

$$c_i(X + e_i \Delta t, t + \Delta t) = c_i(X, t) - \frac{\Delta t}{\tau'}\left[c_i(X, t) - c_{\mathrm{eq}i}(X, t)\right] \tag{11.1.2}$$

where $f_i(X, t)$ and $c_i(X, t)$ are the distribution functions for fluid particles and solute particles in the i direction with a velocity e_i at position X and time t, respectively, where in the two-dimensional nine-speed (D2Q9)model adopted, the nine possible particle velocities (e_0,e_1,...,e_8) are given by

$$[e_0, e_1, e_2, e_3, e_4, e_5, e_6, e_7, e_8] = c\begin{bmatrix} 0 & 1 & 0 & -1 & 0 & 1 & -1 & -1 & 1 \\ 0 & 0 & 1 & 0 & -1 & 1 & 1 & -1 & -1 \end{bmatrix} \tag{11.1.3}$$

Z. Dou et al., *Mass Transfer Dynamics of Contaminants in Fractured Media*,
https://doi.org/10.1007/978-981-99-9187-7_11

Where $c = \frac{\Delta x}{\Delta t}$, with Δx being the lattice space and Δt being the time step; τ and τ' are two different non-dimensional relaxation times; $f_{\mathrm{eq}i}(X,t)$and $c_{\mathrm{eq}i}(X,t)$are the equilibrium distribution functions for fluid and solute, respectively, and are defined as,

$$f_{\mathrm{eq}i}(X, t) = \omega_i \rho \left[1 + \frac{e_i \cdot u_{\mathrm{eq}}}{c_s^2} + \frac{(e_i \cdot u_{\mathrm{eq}})^2}{2c_s^4} - \frac{(u_{\mathrm{eq}})^2}{2c_s^2} \right] \tag{11.1.4}$$

$$c_{\mathrm{eq}i}(X, t) = \omega_i C \left[1 + \frac{e_i \cdot u_{\mathrm{eq}}}{c_s^2} + \frac{(e_i \cdot u_{\mathrm{eq}})^2}{2c_s^4} - \frac{(u_{\mathrm{eq}})^2}{2c_s^2} \right] \tag{11.1.5}$$

where c_s is the lattice sound speed (Lu/ts); u_{eq} is the local equilibrium velocity vector; ω_i is the weight, and for the D2Q9 model, is given by

$$\omega_i = \begin{cases} \frac{4}{9}, & i = 0 \\ \frac{1}{9}, & i = 1, 2, 3, 4 \\ \frac{1}{36}, & i = 5, 6, 7, 8 \end{cases} \tag{11.1.6}$$

ρ and C are the macroscopic fluid density and solute concentration, respectively, and can respectively be expressed as,

$$\rho = \sum_{i=0}^{8} f_i \tag{11.1.7}$$

$$p = c_s^2 \rho \tag{11.1.8}$$

The solute concentration C is expressed as,

$$C = \sum_{i=0}^{8} C_i \tag{11.1.9}$$

Depending on the different dimensionless relaxation times (τ and τ') in Eqs. (11.1.1) and (11.1.2), the corresponding kinematic viscosity and diffusion coefficient are obtained by

$$\upsilon = c_s^2 (\tau - 0.5) \Delta t \tag{11.1.10}$$

$$D_f = c_s^2 (\tau' - 0.5) \Delta t \tag{11.1.11}$$

The lattice Boltzmann model for the fluid field can be used to reconstruct the Navier-Stoke equation with second-order accuracy using the Chapman-Enskog expansion [Eq. (11.1.1)]. The macroscopic advection–diffusion equation can also be used to recover the solute particle distribution function [Eq. (11.1.2)]. As a result, the governing equations for the model's flow and concentration fields are as follows:

$$\frac{\partial \rho}{\partial t} + \nabla \cdot (\rho u) = 0 \tag{11.1.12}$$

$$\frac{\partial (\rho u)}{\partial t} + \nabla \cdot (\rho u u) = -\nabla p + \upsilon\{\nabla^2(\rho u) + \nabla[\nabla \cdot (\rho u)]\} \tag{11.1.13}$$

$$\frac{\partial C}{\partial t} + u \cdot \nabla C = \nabla \cdot (\nabla D_{\mathrm{f}} C) \tag{11.1.14}$$

where u is the velocity vector.

Any lattice node in the computational domain in our simulations can be either a solid node or a fluid node. Prior to streaming, the bounce-back algorithm for the solid node is implemented with a non-slip wall boundary condition in place of the collision step. The C++ code is used to implement every simulation.

11.2 Taylor Dispersion Simulation Based on LBM

The simulation of Taylor dispersion (see Fig. 11.1) provides a straightforward illustration of the possibilities of LBM for solute transport in a rough fracture. The simplified multi-component Shan-Chen model, according to Sukop and Thorne (2005), can mimic Taylor dispersion. The lattice-gas model was used by Stockman et al. (1997) to simulate Taylor dispersion:

$$D = D_{\mathrm{m}} + \frac{b^2 v^2}{210 D_{\mathrm{m}}} \tag{11.2.1}$$

where D is the dispersion coefficient, b is the aperture within a single fracture, D_{m} is the molecule diffusion coefficient, and v is the mean velocity of the flow. Taylor dispersion occurs as a result of the velocity variations across the fracture aperture

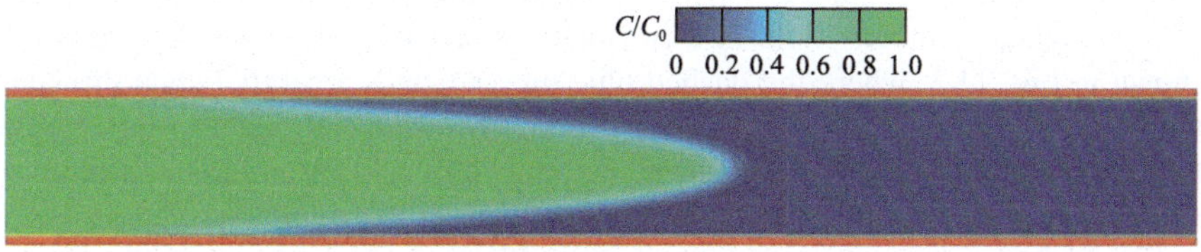

Fig. 11.1 Taylor dispersion in a single fracture

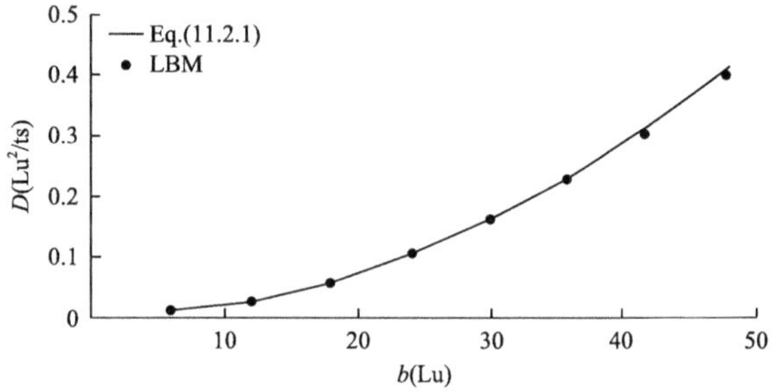

Fig. 11.2 Dispersion coefficients obtained from LBM and Eq. (11.2.1)

due to diffusion. Here we use two regular lattice Boltzmann equations to demonstrate Taylor dispersion. Figure 11.1 shows Taylor dispersion in a single fracture, where the black contours represent the fracture wall surfaces and C_0 is the initial solute concentration.

In order to determine if LBM can accurately represent the Taylor dispersion in a single fracture, a series of case studies with various apertures are done. The channel lengths range from 100 to 1000 Lu, with widths ranging from 5 to 50 Lu. By modifying the external body force, the average velocities for all simulations are set to 0.007 Lu/ts. $D_m = 0.0013$ Lu^2/ts is used as the diffusion coefficient for molecules. The analytical response to Equation [Eq. (11.2.1)] is shown in Fig. 11.2. Applying the calculated breakthrough curves using LBM to the solutions of the advection–dispersion equation, and the predicted dispersion coefficient (ADE). It is evident that the LBM results agree with the computational outcome.

11.3 Characteristics of Solute Transport in a Single Rough Fracture

The continuous injection approach is used to run a number of simulations of solute transport in a single rough fracture at various fluid velocities. For the upper and lower fracture walls, respectively, two self-affine fracture walls with $H = 0.8$ and $H = 0.7$ are created. A two-dimensional variable aperture fracture is represented by a computational domain of 550×130 Lu. The mobility is in the left-to-right direction (see Fig. 11.3). The intake and outflow boundary conditions for a concentration field are, respectively, full developed and constant concentration. The periodic boundary condition is applied for the flow field's entrance and outlet, respectively. The steady-state velocity field in a two-dimensional variable- aperture fracture is shown in Fig. 11.3, where the dashed blue line indicates a parallel single fracture with a mean aperture $\overline{b} = 63\,Lu$, the black arrow with tail denotes the velocity field, and the red line denotes the fracture wall. The velocity field is non-uniform and changes with the fracture's apertures, as seen in the figure. As demonstrated in Fig. 11.3, the mean aperture $\overline{b}$ for the variable-aperture fracture is 63 Lu due to

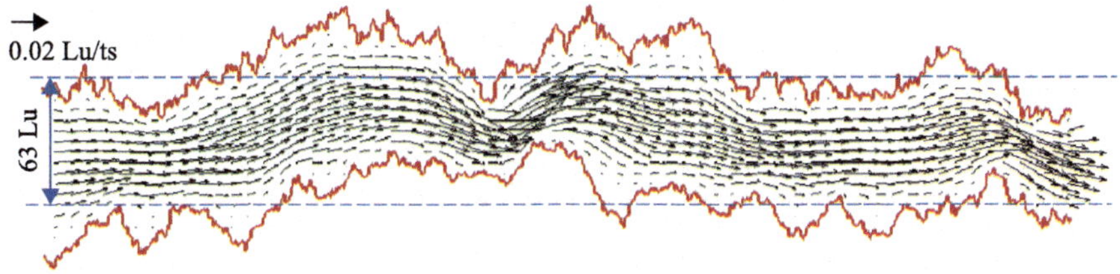

Fig. 11.3 Flow field in a single rough fracture

computational capacity limitations. It is recognized that the mean aperture significantly affects the transport of solutes. For the approaching single fractures, the mean aperture is therefore constant. The following is the traditional analytical solution with the longitude dispersion coefficient DL for the smooth parallel fracture with continuous injection of a constant concentration (Bodin et al. 2003):

$$C_{\mathrm{f}}(x,t)=\frac{C_0}{2}\left[\operatorname{erfc}\left(\frac{x-vt}{2\sqrt{D_{\mathrm{L}}t}}\right)+\exp\left(\frac{vx}{D_{\mathrm{L}}}\right)\operatorname{erfc}\left(\frac{x+vt}{2\sqrt{D_{\mathrm{L}}t}}\right)\right] \quad (11.3.1)$$

with the following initial and boundary conditions:

$$C_{\mathrm{f}}(x,0)=0 \quad (11.3.2)$$

$$C_{\mathrm{f}}(0,t)=C_0 \quad (11.3.3)$$

$$C_{\mathrm{f}}(\infty,t)=0 \quad (11.3.4)$$

where $C_{\mathrm{f}}(x,t)$ is the solute concentration at position x and time t.

By altering the external body force, two examples with different *Re* values (*Re* = 13 and *Re* = 86) are explored in to investigate the effect of roughness on solute transport at various fluid velocities.

Figure 11.4 depicts the solute transport procedure in the rough fracture with *Re* = 13, with the fracture wall being represented by the white contour. As observed in the figure, the solute transport channel is significantly impacted by the roughness of a fracture wall. Figure 11.3 demonstrates how the distribution of the aperture has a significant impact on the flow field in the rough fracture. Due to the comparatively high local fluid velocity, solutes flow through the aperture preferentially in the center section. As a result of the roughness, the solute fronts are first noticed near the centre of the aperture, and the front spreading pathway changes with the local fluid velocity. As a result, the mobile region for solute transport is significantly impacted by the fracture geometry. The dominating molecular diffusion causes the concentration of the solute front to decrease when more solutes are injected. This could lead to a quicker breakthrough time. Due to the transverse concentration gradient, once solutes have penetrated the entire fracture, they continue to diffuse in the direction of

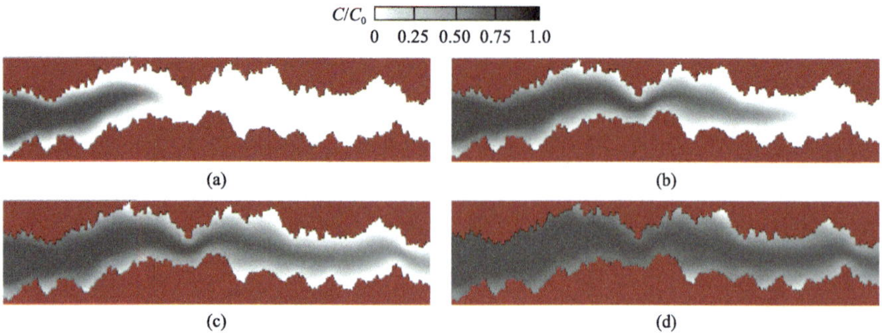

Fig. 11.4 Process of solute transport in the rough fracture with $Re = 13$, **a** $t = 14{,}500$ ts; **b** $t = 22{,}000$ ts; **c** $t = 32{,}000$ ts; **d** $t = 60{,}000$ ts

the rough fracture walls [comparing Fig. 11.4c, d]. This difference in concentration causes a transverse flux of solutes, which improves macro-dispersion. Nonetheless, some areas with little solute (immobile regions) are seen (Fig. 11.4d) and continue to have low concentrations. On the one hand, the small solute flow and nearly zero fluid velocities in the region are the causes of these immobile regions for solute transport. On the other hand, the solute molecule diffusion causes the concentration to rise gradually. The molecular diffusion process regulates the solute transport in the immobile area. The solute molecules should diffuse during the injection time and reach equilibrium at an infinite injection time, according to theory. Nonetheless, it is presumed that the final steady state for solute transport has occurred if the variation in outlet concentration between 8000 ts is less than 0.1%. In fact, even with a tiny *Re*, the equilibrium of solute molecule diffusion in these areas will persist for a very long period in a field test. In Fig. 11.4d, the movable and immobile zones are thus formed.

The solute transport procedure in the rough fracture with $Re = 86$ is depicted in Fig. 11.5. The rough fracture's aperture distribution in this instance is identical to that in the case with $Re = 13$. Figure 11.5 demonstrates how the solute transport in the rough fracture is significantly impacted by the rising fluid velocity. The predominant advection mechanism causes a rise in the gradient of solute concentration. The center of the aperture, where the fluid velocity is quite high, is where the solute front with the highest concentration gradient expands. Despite the fact that advection dominates solute transport, solutes can diffuse in the direction of the rough fracture wall, where the concentration of solutes is relatively low. The increase in marco-dispersion caused by this delayed diffusion process is comparable to the situation where $Re = 13$. On how the solute concentration is distributed, the flow field has a big impact. One thing that becomes clear when comparing Figs. 11.4d and 11.5d is the distinction between the stationary and movable sections. This is true because a stronger advection effect on solute transport is associated with a higher Re value. Due to the ultimate steady state's conditions, the sluggish process of solute molecule diffusion only requires a short amount of time. Yet, it is discovered that the immobile region's area is larger than it was when $Re = 13$. Furthermore susceptible to the fracture shape is the

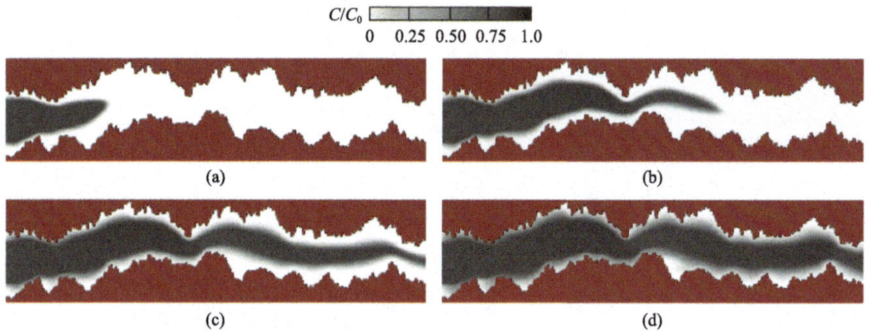

Fig. 11.5 Process of solute transport in the rough fracture with $Re = 86$, **a** $t = 10{,}000$ ts; **b** $t = 16{,}000$ ts; **c** $t = 21{,}000$ ts; **d** $t = 45{,}000$ ts

immobile zone along the rough fracture wall. The smaller immobile zone for solute transport in the rough fracture is caused by the reduced roughness.

Figure 11.6 displays the BTCs for various *Re* values in the rough and smooth fractures, where the dimensionless time $T = t/T_0$, where t is the time step and T_0 is the total time step. The equivalent *Re* value for the solute transport in the smooth fracture is consistent with that in the case when $Re = 13$ in the rough fracture. Equation (11.3.1) determines the analytical solution curve. The LBM findings and the analytical answers accord fairly well. The largest difference between analytical answers and LBM results is less than 0.15%. Figure 11.6 demonstrates that BTCs are significantly impacted by both *Re* and roughness. The lengthy breakthrough time for the BTC denotes a lengthy travel period. The longer travel time in the rough fracture is caused, as would be predicted, by the smaller *Re*. Moreover, compared to the smooth fracture with $Re = 13$, the travel time for cases with both $Re = 13$ and $Re = 86$ is less. The fluid velocity at the center of the aperture for the rough fracture has a significant impact on the solute molecule's journey time. *Re* is the same for both smooth and rough fractures, but the travel duration varies. This is because the variation in the aperture in the rough fracture causes the fluid velocity in the middle of the aperture for the rough fracture to be higher than that for the smooth fracture. The roughness affects the trip time.

For situations of the rough fracture, lengthy tails on BTCs are also seen at the end. According to Raven et al. (1988), the diffusive interchange of solutes between the mobile and immobile areas accounts for the extended breakthrough tail in the case of the rough fracture. The dimensionless concentration with $Re = 86$ is higher than that with $Re = 13$, as the larger mobile region is caused by the greater *Re*.

The solute transport in a rough fracture may be studied extremely effectively using the LBM, as this chapter demonstrates. The paths of solute transport are divided into the mobile and immobile zones because roughness causes the non-uniform fluid velocity field in the fracture. In the rough fractures, the mobile region is dispersed along the central aperture, and the advection mechanism dominates for solute transport. Along the rough fracture walls, an immobile zone is visible, and the solute

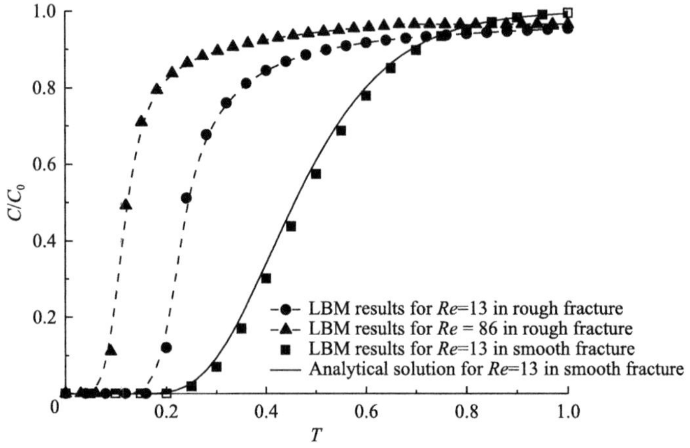

Fig. 11.6 Breakthrough curves for different *Re* values and types of the fracture

molecule diffusion process is predominant. When *Re* rises, the mobile region expands along with the front of high solute concentration. The immobile area becomes more pronounced as both *Re* and fracture geometry are increased. A lengthy breakthrough tailing is also a result of the slow solute molecule exchange between the mobile and stationary areas. In addition to *Re*, roughness also affects the concentration of solutes in the front region. The number of solute fronts grows as *Re* increases. Because molecular diffusion is the primary process, the concentration of the solute fronts drops. This could lead to a quicker breakthrough time.

References

Bodin J, Delay F, de Marsily G (2003) Solute transport in a single fracture with negligible matrix permeability: 2. mathematical formalism. Hydrogeol J 11(4):434–454

Raven KG, Novakowski KS, Lapcevic PA (1988) Interpretation of field tracer tests of a single fracture using a transient solute storage model. Water Resour Res 24(12):2019–2032

Stockman HW, Li C, Wilson JL (1997) A lattice-gas and lattice Boltzmann study of mixing at continuous fracture junctions: importance of boundary conditions. Geophys Res Lett 24(12):1515–1518

Sukop MC, Thorne DT (2005) Lattice boltzmann modeling: an introduction for geoscientists and engineers. Springer, New York

GPSR Compliance

The European Union's (EU) General Product Safety Regulation (GPSR) is a set of rules that requires consumer products to be safe and our obligations to ensure this.

If you have any concerns about our products, you can contact us on ProductSafety@springernature.com

In case Publisher is established outside the EU, the EU authorized representative is:

Springer Nature Customer Service Center GmbH
Europaplatz 3
69115 Heidelberg, Germany

The manufacturer's authorised representative in the EU is Springer Nature Customer Service Centre GmbH, Europaplatz 3, 69115 Heidelberg, Germany. If you have any concerns regarding our products, please contact ProductSafety@springernature.com

Printed and bound by CPI Group (UK) Ltd, Croydon, CR0 4YY
01/10/2026
02225868-0001